HOME
CHEESE
MAKING

From Fresh and Soft to Firm,
Blue, Goat's Milk, and More

Ricki Carroll

Additional recipe development by Jim Wallace

Storey Publishing

*The mission of Storey Publishing is to serve our customers by
publishing practical information that encourages
personal independence in harmony with the environment.*

Edited by Lisa Hiley, Evelyn Battaglia, and Alana Chernila
Technical assistance by Mark Chrabascz
Art direction and book design by Alethea Morrison
Text production by Jennifer Jepson Smith
Indexed by Christine R. Lindemer, Boston Road
    Communications

Cover and interior photography by © Evi Abeler Photography
Additional photography by Courtesy of Bellwether Farms, 80;
    © blickwinkel/Alamy Stock Photo, 6 right; Brian Birzer,
    287; © Brian Jannsen/Alamy Stock Photo, 4; © DEA/G.
    DAGLI ORTI/Getty Images, xiii; Courtesy of Deb Hahn,
    223; © Gideon Mendel/Getty Images, xii; Hemant,
    85; Imran Saleh, 94; © Jeremy Lago, 276; © Jonathan
    Bielaski, 58; © Jonathan Irish/Getty Images, xiv; Josh
    Karp, 139; © LOOK Die Bildagentur der Fotografen
    GmbH/Alamy Stock Photo, 118; Courtesy of Luke Padgett,
    167; Mars Vilaubi, 35 bottom, 270; © Meredith McKown
    Photography, 247; Muwonge Baker (Managing Director),
    114; Courtesy of Nettle Meadow, 133; © Point Reyes
    Farmstead Cheese Company, 242; Courtesy of Red Barn
    Family Farms, 219; Courtesy of Ruth Appel, 90; © Stan
    Pritchard/Alamy Stock Photo, 5; Susan Sellew, 252;
    © Zoonar GmbH/Alamy Stock Photo, 6 left

Food styling by Alana Chernila and Joy Howard
Prop styling by Christina Lane
Illustrations by © John Burgoyne

Text © 1982, 1996, 2002, 2018 by Ricki Carroll
Home Cheese Making was first published in 1982 as
Cheesemaking Made Easy and was revised in 1996. For this
edition, all of the information in the previous editions was
reviewed and updated, and new recipes were added.

Some of the recipes in this book have been adapted from recipes
developed for cheesemaking.com by our technical advisor and
cheese guru, Jim Wallace.

Be sure to read all instructions thoroughly before using
any of the techniques or recipes in this book and follow
all safety guidelines.

Storey Publishing
210 MASS MoCA Way
North Adams, MA 01247
storey.com

Printed in the United States by Versa Press
10 9 8 7 6 5 4 3 2 1

LIBRARY OF CONGRESS CATALOGING-IN-PUBLICATION DATA
Names: Carroll, Ricki, author.
Title: Home cheese making : from fresh and soft to firm, blue,
    goat's milk, and more : recipes for 100 favorite cheeses /
    by Ricki Carroll.
Description: 4th edition. | North Adams, MA : Storey
    Publishing, 2018. | Includes index.
Identifiers: LCCN 2018030470 (print)
    | LCCN 2018033260 (ebook)
    | ISBN 9781612128689 (ebook)
    | ISBN 9781635860788 (hardcover : alk. paper)
    | ISBN 9781612128672 (pbk. : alk. paper)
Subjects: LCSH: Cheesemaking. | Cooking (Cheese)
    | LCGFT: Cookbooks.
Classification: LCC SF271 (ebook)
    | LCC SF271 .C37 2018 (print)
    | DDC 637/.3—dc23
LC record available at https://lccn.loc.gov/2018030470

CONTENTS

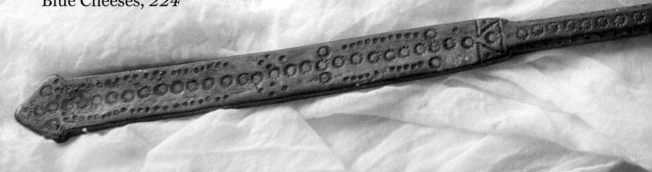

Foreword

Two hundred years ago, all American cheese was made in the home. In 1800, nearly 90 percent of the country's population lived on farms, and making cheese was considered the work of the farm wife, just part of the routine of baking and cooking and other chores. No need for a how-to book like this — cheese making was something every eight-year-old was probably familiar with.

The first cheese factory was begun by Jesse Williams in 1851, in Rome, New York. Food-producing factories were hailed as a liberating force for farm women, releasing them from the demand of daily chores. Eventually people mostly "forgot" how to make cheese. This book came into being in the 1980s in response to an increasing interest in those old homesteading skills.

Ricki's wonderful work allows modern-day home cheese makers access to the dairy wisdom of centuries past. It also gives us something our ancestors didn't have — the ability to make dozens of different cheeses in our own kitchen. Two hundred years ago, when you made cheese at home, the odds were high that it would have been the same cheese your mother made. The same one her mother made, and the same one your neighbor made. Knowledge of other cheeses would have been extremely limited. With recipes representing nearly every cheese-making region, *Home Cheese Making* puts a world of homemade cheese at your fingertips.

To be clear, as the owner of a business that makes and sells cheese to food lovers all across the country, it's not like I want you to make *all* your own cheese. But I strongly encourage you to turn back the culinary clock and start making at least *some* of your own cheese, the way so many American families did two hundred years ago. I think you'll find that making homemade cottage cheese or 30-Minute Mozzarella creates a family bond. The sense of achievement, the fun of eating something you made yourselves, and the pleasure of fresh cheese coming to life on your stovetop are pretty magical.

Thanks to Ricki's diligent recipe testing and nearly 40 years of teaching cheese-making classes, the recipes here really work! With *Home Cheese Making* in hand, the odds of you making great cheese are very high. Ricki walks you through each step in easy-to-understand, layperson's language.

For your own sake and for the sake of the country, I really hope you buy — *and use* — *Home Cheese Making.* I forecast that within a few months you'll find yourself starting to serve homemade cheese at holidays throughout the year, making it with your kids for their school lunches, and giving your cheese to friends and loved ones as gifts. Your world will be a tastier, calmer, and more caring place for it.

Ari Weinzweig
cofounder of Zingerman's Community of Businesses
and former president of The American Cheese Society

THE ART OF CHEESE MAKING

Those of us who remember our first attempts at bread making look back with an indulgent smile upon the bowl of sticky, disobedient dough that clung to everything it touched and defied our inexperienced hands at every turn. The smile is possible because we also remember when we finally beat the dough into submission and it did everything the cookbook promised it would. We kneaded it into a wonderful elastic mass that rose up gloriously under its dampened cloth cover and soon filled the entire house with that heavenly smell.

No one seems to know why the art of bread making fled the factories and settled with delicious comfort in our homes so far ahead of the art of cheese making. Bread, wine, and cheese, besides being (collectively and individually) among life's greatest pleasures, share another bond: They are all produced by the fermentation process. Bread rises in the pan, and wine matures through the reaction of yeast and sugar. Cheese, as if by magic, emerges from a pot of milk through the action of bacteria and lactose (milk sugar).

PRESERVING THE PAST

This book is designed to keep alive the art and enjoyment of home cheese making. Just as there's a knack to handling bread dough, there are tricks to turning milk into cheese. You may have learned to make bread by watching your mother or grandmother. Today, there are more and more mothers and grandmothers — as well as aunts, uncles, siblings, and friends — who can be counted on to hand down the techniques of cheese making to the next generation.

If you follow the directions carefully (especially the ones about cleanliness), I can almost guarantee success — and the undeniable thrill of watching your friends smack their lips over slices of your very own Gouda while you say, "I made it myself!" But on the off chance the going gets rough, the curds won't set, or your cheese turns out less than perfect, you can refer to the troubleshooting section starting on page 351. And there is always help available to guide you at www.cheesemaking.com. Keeping good records is helpful to you and will enable us to help you better (see an example on page 355).

CHEESE, THE ORIGINAL SLOW FOOD

In our fast-paced world, any organization whose official symbol is the snail must have its heart in the right place. Slow Food is an international movement that started in Italy in 1986 as a protest against McDonald's bringing its golden arches into historic Rome.

But Slow Food isn't just against change: It is a growing voice in favor of family farms, sustainable agriculture, biodiversity, eating seasonally and regionally, and many other laudable goals. Most of all, Slow Food's message is this: Food should taste good. That means using fresh ingredients, careful preparation, and respect for recipes and traditions lasting for generations. "Slow food" often is simple food. It tastes best eaten at a table with friends and family.

Cheese may be the original slow food. Slow is the only way to go if you make cheese, wine, beer, bread, or even pickles, which all require age-old processes that cannot be rushed. Slow Food appreciates the care home and farmstead cheese makers take in recalling the old alchemy of transforming milk into cheese every time you slowly warm milk, gently stir the curds, and patiently wait for your cheese to ripen to perfection.

MADE BY JOE SCHNEIDER USING RAW MILK DIRECT FROM THE FARM

IN THE BEGINNING . . .

Cheese is one of the earliest foods made that people still consume. It dates back to the initial domestication of animals, estimated to at least 9000 BCE. Archaeologists have established that cheese was well known to the Sumerians (4000 BCE), whose cuneiform tablets contain references to cheese, as do many Egyptian and Chaldean artifacts. It was a staple in biblical times, along with honey, almonds, and wine, and is associated with stories of great daring: David was delivering cheese to Saul's camp when he encountered Goliath.

It's fun to speculate about how cheese may have been discovered. I like the oft-told tale about the nomad who poured his noon ration of milk into a pouch made from the stomach of a sheep or camel (just the right size for traveling and lined with enigmatic enzymes) and plodded across the hot desert all morning, only to find, at noon, that his liquid lunch had solidified! It must have been a shock for the poor wanderer. He may not have been as brave as the person who ate the first raw oyster, but I imagine our hero eating a rather late lunch that day.

Once he finally tasted it, however, he probably became the first traveling salesman, singing the praises of this delicious new food at all his overnight oasis stops. I suppose it didn't take long for his customers to recognize a good thing when they tasted it and for the cheese-making mystique to take hold and spread around the world.

HISTORY IN THE MAKING

Cheese making was important to the ancient Greeks, whose deity Aristaeus, a son of Apollo, was considered the giver of cheese. Homer sang of cheese in the *Odyssey*. A Greek historian named Xenophon, born circa 431 BCE, wrote about a goat cheese that had been known for centuries in Peloponnesus. And the original Greek Olympic athletes trained on a diet consisting mostly of cheese.

The craft of cheese making was then carried westward to Rome. The Romans refined techniques, added herbs and spices, and invented a method for smoking cheese. Many varieties were made during those times, and the Romans feasted on curd cheeses, Limburger-type cheeses, soft cheeses, and smoked and salted cheeses. They exported their hard cheeses and experimented with a cheese made from a mixture of sheep's and goat's milk.

In addition, these resourceful Romans — among the first *formaggiaio* — learned to use different curdling agents besides the rennet they extracted from the stomach of a weanling goat or sheep. Thistle flowers, safflower seeds, and fig bark

Terra-cotta figurine of a man grating cheese, sixth century BCE, Greece

were soaked in water to make extracts to set a curd. They devised baskets, nets, and molds to shape their cheeses. Indeed, when Julius Caesar sent his soldiers to conquer Gaul, they packed rations of cheese for the long marches; wherever the legions went, northern tribesmen quickly learned to copy their captors' delicious food.

FROM FIELDS TO FACTORIES

In the fifteenth century, far up in the Alps, Swiss farmers milked their cows in the fields and brought the milk back to their farms to make cheese. It wasn't until about 1800 that they realized they could make cheese down in the valleys as well as high in the hills. In 1815, the first cheese factory was opened in the valley at Bern. It was such a successful venture that 120 cheese factories sprouted in the next 25 years, with the number growing to 750 by the end of the nineteenth century.

The early Swiss factories had a fire pit in the corner, with a copper kettle hanging over it on a crane so they could swing the kettle above or away from the fire. (This technique is still used in a handful of rural places around the world.) They tested the temperature on their forearms. Their rennet solution, made from the stomach lining of a calf, was so strong they claimed it could set milk in the time it took to recite the Lord's Prayer. (This is an interesting idea, but when I tried it, even slowly, I could only get it to one minute, a trifle short for the setting time in most of these recipes.)

The first American cheese factory was established in 1851 in Rome, New York, by an ingenious entrepreneur named Jesse Williams. Earlier companies, as far west as Ohio, had been set up to buy homemade curds from local farmers and process the curds into cheeses weighing from 10 to 25 pounds. Williams realized that cheese made from several different batches of curd lacked uniformity in taste and texture. He also knew that it takes exactly the same length of time to make a curd from 1,000 pounds of milk as it does from one pint, so he set up his factory to make cheese from scratch.

He bought milk from local farmers to add to the output of his own herd of cows, and in its first years of operation, his factory produced four cheeses per day, each weighing at least 150 pounds. Springwater was circulated around the vats to cool them, and steam, produced by a wood-fired boiler, heated them. It is recorded that his costs averaged about 5½ cents a pound for an aged cheese and that he sold this entire stock on contract for a minimum of 7½ cents a pound. The sale of pork from his large herd of hogs, which were fed the factory's whey output, enhanced the financial success of the operation. (I think he might have been one of the first artisanal cheese makers in America.)

Making cheese in a traditional Swiss copper vat

From the Latin *caseus* for "cheese" came the German *Käse*, the Dutch *kaas*, the Irish *cuis*, the Welsh *caws*, the Portuguese *queijo*, and the Spanish *queso*. The Anglo-Saxon *cese*, or *cyse*, also became *cheese*.

Scientific methods developed in the nineteenth century took the guesswork out of commercial cheese making. Models of cleanliness and efficiency, cheese factories are equipped with steam-jacketed stainless-steel vats, thermostatic controls, and mechanical agitators and curd cutters. Sterile conditions largely eliminate the possibility of unwanted bacterial invasion at any stage of a cheese's development.

When I started making cheese from our goats in the early 1970s, I was part of a burgeoning do-it-yourself culture. In addition to our concern for the environment, as well as having more milk than we could drink, we were trying to reclaim some of the techniques of self-sufficiency that made our ancestors totally independent. We learned how to heat with wood, bake bread, make dandelion wine, grow soybeans and make tofu, and of course, make our own cheese. Since then, an entire artisanal cheese movement has grown up in the United States and continues to gain momentum as new generations embrace a back-to-basics approach to what we put on our plates.

When you learn what it takes to make a delicious product, you will be drawn to buy more nutritious and delicious products. If you know how to make bread, you go to the bakery and buy the most delicious breads you can find. As you learn to make cheese, you will gain a deeper appreciation of the artisans who produce some of the most amazing cheeses in the country. Buy their cheeses, try new kinds, and support the makers — they are helping to save our quickly disappearing farmland and they need our help to survive. Do what you can, when you can, and thank your local artisanal cheese makers for all their extraordinary work.

HOW TO USE THIS BOOK

The cheese recipes in this book are divided into common categories: fresh, soft and semi-soft mold-ripened, hard and semi-hard, blue, goat, and whey cheeses. There are recipes for cultured dairy products as well. Each chapter includes a variety of cheeses made in a similar fashion and involving the same basic steps and techniques.

Just as we must all walk before we can run, it is smart to start cheese making with the simpler varieties. Begin by making a few of the fresh cheese recipes before you graduate to the complexities of the hard cheeses. The fresh cheeses I recommend for beginners are Queso Blanco, Fromage Blanc, and Whole-Milk Ricotta; 30-Minute Mozzarella is also one of the most fun cheeses to make if you have good milk (all are found in chapter 4). For your first aged cheese, try Feta (also in chapter 4), or Farmhouse Cheddar or Gouda (found in chapter 6). Always keep accurate records to learn from your mistakes and benefit from your successes. Read on, and happy cheese making.

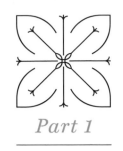

Part 1

GETTING STARTED

INGREDIENTS

When early humans first discovered that goats, sheep, cows, and other mammals produced milk, they soon realized they could use the wondrous, wholesome substance to feed their families. The further discovery that milk stored in pouches made of animal stomachs coagulated and turned into a nutritious food was nothing short of a miracle. Stomach linings became a source of rennet, and soured milk and whey became a source for cultures. The miracle of cheese solved the age-old question of how to save precious, perishable milk. The discovery of naturally occurring molds added a vim and vigor to cheese and introduced a delectable variety to the palate, leading to an insatiable drive to develop more and more styles of cheeses.

Today, we use the same ingredients, although they are usually obtained in more convenient ways. Our milk may come in cartons from the grocery store (although more and more people are buying it locally from farmers). Cultures, molds, and rennets — vegan versions, too — are now standardized in factories and can be obtained from cheese-making supply houses. But hark! I hear the artists calling, because with the right techniques, with these ingredients, you can turn milk into gastronomic delights.

1

DECODING MILK LABELS

I completely support the use of organic milk; unfortunately many supermarket varieties are ultrapasteurized and therefore not conducive to cheese making. (See Pasteurization to Ultrapasteurization, page 11.) If you are able to find a local source for organic milk, use it. There are also some online sources of organic milk that are not ultrapasteurized. Here is a little lesson in the definition of "organic" and other common food labels for milk and dairy products.

Organic: Organic milk, by regulation, is supposed to come from a cow whose milk production was not prompted by an artificial growth hormone, whose feed was not grown with pesticides, and who had "access to pasture." That last part has long raised concern among consumer groups who claimed some producers were ignoring the original intent of the regulations by doing just enough to meet the letter of the law, with the cows spending most of their milk-producing lives confined to cramped feed lots eating grain and not grass. In 2010, the regulation was clarified to specify cows must spend a minimum of 120 days outside during the growing season, although the question remains if and how the requirement is being enforced.

Many farmers are going to great lengths to provide organic milk for their lucky customers. Sadly, the vast majority of organic milk comes from the same kind of factories where conventional milk is mass produced, using the same breeding and feeding methods — and of course the milk is ultrapasteurized, making it difficult to use for cheese making.

Certified Humane: This is the label to look for if you want milk produced from cows who were pasture-raised and allowed freedom of movement as well as being free from artificial growth hormones.

Animal Welfare Approved: If you see the AWA label on milk, the cows must have access to the outdoors and be able to engage in natural behavior. As with Certified Humane, AWA milk is free of artificial growth hormones.

Hormone-Free, rBGH-Free, or rBST-Free: Genetically engineered hormones such as rBGH and rBST are commonly used to increase milk production and are associated with animal welfare concerns. Milk bearing these labels (or simply "no hormones added") will be free of those hormones but there is no indication of how the cows were raised.

Milk

Fresh milk from a healthy animal is about as good as it gets, containing its own system of cultures and enzymes that makes it exactly right for cheese making. While you may think of it as the white liquid in the plastic jugs, milk is quite a complicated substance. About seven-eighths of it is water. The rest is made up of proteins, minerals, milk sugar (lactose), milk fat (butterfat), vitamins, and trace elements. Those substances are called milk solids.

When we make cheese, we cause the protein part of the milk solids, called casein, to coagulate (curdle) and produce curd. At first the curd is a soft, solid gel, because it still contains all the water along with the solids. As it is heated over varying lengths of time, the curd releases liquid (whey), condensing more and more of the solids until it becomes cheese. Most of the butterfat remains in the curd and very little passes into the whey. Time, temperature, and a variety of friendly bacteria determine the flavor and texture of each type of cheese.

Throughout history, people have used milk from many animals. The familiar cow, goat, and sheep have fed people for centuries, along with less common animals such as the yak, camel, water buffalo, llama, ass, elk, mare, caribou, zebu, and reindeer. When making the cheeses in this book, you may use whatever milk you have available in your area. Cow's milk and goat's milk are the most readily available in the United States; you may also find some sheep's milk and water buffalo's milk, if you are extremely lucky.

You can make most of the cheeses in this book with store-bought milk, as long as it has not been ruined by high-temperature pasteurization techniques (see page 11). With a few specified exceptions, you may use whole, skim, or any percent of milk in these recipes — it has always been my mission to render cheese making accessible to everyone.

Whole milk gives a higher yield and tastes best because of the butterfat it contains, whereas skim milk gives you a drier cheese and a lower yield. Experimenting will help you find your favorite result. Dried milk powder can often be used for the recipes in chapters 4 and 5 (Fresh Cheeses and Soft and Semi-Soft Ripened Cheeses, respectively) and in chapter 10 (Cultured Dairy Products).

Fresh, local milk will give your cheeses the most complex flavors. No matter what type of milk you

use for cheese making, it must be of the highest quality. Use the freshest milk, from a local brand whenever possible. Long-term cold storage and transportation can be problematic for milk, so the closer the milk is to the source (be it cow, goat, or sheep), the better for using it to make cheese.

If your milk comes from the supermarket, do not open the container until you are ready to start your cheese making. This will prevent possible contamination by bacteria in the air. Above all, if the milk tastes sour or "off," throw it out — the cheese-making process will not make spoiled milk taste better.

When purchasing cow's milk, remember one gallon yields approximately one pound for hard cheeses and approximately two pounds for fresh and soft cheeses. This varies from type of milk and breed of animal. Yields from goat's milk tend to be slightly lower due to the butterfat content, and low-fat milk will give the lowest yield and a drier cheese. The yield from sheep's milk will be highest; however, sheep give very little milk compared to cows.

COW'S MILK

In the United States today, cow's milk is most likely to be used in cheese making. However, goats and sheep feed the majority of the world's population. Cows are large animals and more difficult to raise; they eat more and take up more grazing space and natural resources. Yet cow's milk is abundant, the curd is firm and easy to work with, and it produces delicious cheese.

Milk from most breeds will work nicely for making any cheese, although generally speaking, certain breeds are better suited for specific types of cheeses. The differences are based on the size and

AVERAGE COMPOSITION OF COW'S MILK

Water	87.0%
Albuminous protein	0.4%
Casein	3.3%
Lactose	4.8%
Butterfat	3.8%
Minerals	0.7%
Total solids	13.0%

amount of butterfat globules in the milk they produce. Jersey and Guernsey milk has the largest fat globules, making it perfectly suited for fresh, soft, and semi-soft cheeses. Ayrshire milk has the smallest fat globules, so it is preferred for sharp Italian cheeses and long-aged cheddars. Holstein milk is the standard, so it can be used across the board.

Average Composition of Goat's Milk

Water	86.0%
Albuminous protein	1.0%
Casein	3.3%
Lactose	4.4%
Butterfat	4.5%
Minerals	0.8%
Total solids	14.0%

If you are considering buying your own cow, start with a Jersey — her rich milk will produce a high cheese yield because it has a high butterfat content, and Jerseys tend to be very sweet animals.

GOAT'S MILK

Goat's milk has smaller butterfat globules than cow's milk, making it more digestible. It is more acidic than cow's milk, so it ripens faster, and it produces a whiter cheese due to a lack of carotene. Because of its natural homogenization, goat's milk makes a slightly softer cheese than cow's milk, though the butterfat content is about the same. Cheese made from raw goat's milk has a distinct peppery hot pungency caused by naturally occurring lipase enzymes and fatty acids. During the renneting process, you may lower the temperature five degrees in a recipe, because goat's-milk curd tends to be more delicate. Remember to treat these softer curds very gently.

If you are looking for your own goats, Nubians and Alpines are good producers and they tend to have the sweetest milk. Saanens often produce a higher yield of milk with a slightly stronger flavor. Toggenburgs produce a slightly lower yield, often also with a stronger flavor.

Freezing Milk

Many people wonder whether freezing milk has an effect on cheese making. Cow's milk does not freeze well, because the cream separates after freezing and thawing. I have found that goat's and sheep's milk can be frozen up to 30 days and then used for drinking. I do not recommend using frozen milk to make any type of cheese.

SHEEP'S MILK

Sheep's milk is one of the most nutritionally valuable foods available. It is high in protein and vitamins, which so often have to be artificially added to our diet. Sheep's milk contains almost 10 percent less water than cow's or goat's milk and is almost twice as high in solids as cow's milk; therefore, it produces a very high cheese yield — almost 2½ times what you would expect from cow's or goat's milk.

Dairy sheep, though still relatively uncommon in the United States, are gaining a foothold across the country. If you are lucky enough to have access to sheep's milk, note these differences. When adding rennet, use one-third to one-fifth of the amount used for cow's milk, and top-stir carefully (see page 40). When cutting the curd, make larger cubes; when ladling, take thicker slices, or you will lose too much butterfat and the cheese will be too

AVERAGE COMPOSITION OF SHEEP'S MILK

Water	81.1%
Albuminous protein	1.1%
Casein	4.6%
Lactose	4.7%
Butterfat	7.5%
Minerals	1.0%
Total solids	18.9%

AVERAGE COMPOSITION OF WATER BUFFALO'S MILK

Water	82.7%
Albuminous protein	0.7%
Casein	3.8%
Lactose	4.5%
Butterfat	7.5%
Minerals	0.8%
Total solids	17.3%

dry. Use half the amount of salt called for and exert only light pressure when pressing.

WATER BUFFALO'S MILK

Water buffalo's milk, which is traditionally used to make mozzarella in Italy, has nearly twice as much butterfat as cow's milk. It is also higher in protein and milk solids, making it a good choice for a naturally thick, Greek-style yogurt without even having to strain it. Alas, water buffalos are harder to milk than cows, which is why the milk can be hard to find in the United States. If you can find some, it's worth giving it a try.

Milk Terminology

Several terms need to be defined for you to understand some of the changes in today's milk and to know what is being referred to in this book.

RAW MILK

Raw milk is processed on-site. It is filtered and cooled before use. It is not pasteurized, so it has a higher vitamin content than heat-treated milk. Raw milk showcases the fullness and richness of flavors, and has the added advantage of bringing the subtleties of pasturing and the diet of the animal into your final cheese.

MAKING CHEESE WITH RAW MILK

All of our recipes may be made with raw milk, taking the proper precautions described under Raw Milk (above). It is best to use raw milk within 48 hours of milking (if milking your own animal, wait at least 2 or 3 hours before using it). For cow's milk, if you prefer it partially skimmed, allow the fat to naturally rise to the top (in the refrigerator) and skim off the cream from the top before using the remaining milk for cheese.

Follow the directions in the recipe as given. Once you have experience you can make adjustments as necessary. You may be able to use 25 to 50 percent less culture and you may be able to lower the temperatures three to five degrees. When making cheese with raw milk, it is important to top-stir (see page 40) when you see butterfat rising to the surface. This mixes the butterfat back into the body of the milk.

It is usually not necessary to add calcium chloride when making cheese with raw milk because the calcium molecules have not been affected by pasteurization, homogenization, and long-term cold storage. However, many cheese makers use calcium chloride to compensate for seasonal variations in the composition of their milk. I recommend using it — it can't hurt your cheese. We do not use it with the pasta filata cheeses, as it can prevent stretching.

Raw milk contains natural flora, many of which are very useful in cheese making. It may also contain harmful bacteria, or pathogens, that can produce disease in humans. Pathogens that may be found in milk include *Mycobacterium*, which causes tuberculosis; *Brucella*, which causes brucellosis; and *Salmonella*, which causes salmonellosis.

Taking precautions to avoid foodborne illness is important for everyone, but especially for those most vulnerable to disease — children, the elderly, and people with weakened immune systems.

If you consume raw milk or use raw milk to produce cheese, *you must be absolutely certain that there are no pathogens in the milk*. A good rule to follow is: If in doubt, pasteurize (see below).

When using raw milk, make sure it comes from tested animals and is kept scrupulously clean. Never use milk from an animal suffering from mastitis (inflammation of the udder) or receiving antibiotics, which will destroy the helpful bacteria essential in making cheese. If you make raw-milk cheese for sale, U.S. federal law dictates it must be aged a minimum of 60 days to prevent the development of pathogenic bacteria. That said, raw-milk cheeses are some of the best in the world. If you get your milk from a local farm, make sure they are taking all precautions and testing their milk regularly.

How to Pasteurize Milk at Home

Follow these simple steps to pasteurize your milk at home.

1. Heat the milk to 145°F (63°C) for exactly 30 minutes or 161°F (72°C) for 15 seconds, stirring occasionally to ensure even heating. The temperature and time are important. Too little heat or too short a holding time may not destroy all the pathogens. Too much heat or too long a holding time can destroy the milk protein and result in a soft curd often unfit for making cheese.

2. Once heated, place the pot in a sink filled with ice water. Stir constantly until the temperature drops to 40°F (4°C). Rapid cooling is important to eliminate conditions supporting growth of unwanted bacteria.

3. Store the milk in a sealed container in the refrigerator until ready to use.

HOMOGENIZED MILK

This milk has been heat-treated and pressurized to break up the butterfat globules into very small particles so they are distributed evenly throughout the milk and do not rise to the top. Homogenized milk produces a smoother curd, which is also not as firm as the curd from raw milk, so use calcium chloride to make cheese with any homogenized milk.

CREAM-LINE MILK

So named for the "line" separating the cream on top from the milk below, this milk is pasteurized but not homogenized. (And it's delicious! You may remember the glass bottles delivered to the door — that was cream-line milk.)

PASTEURIZED MILK

This type of milk has been heat-treated to destroy pathogens. In effect, pasteurization kills all bacteria. However it also makes proteins, vitamins, and milk sugars less available, and it destroys the enzymes that help the body assimilate them.

ULTRAPASTEURIZED (UP) MILK

Scientists working for large corporations have figured out that if you heat-treat milk to ultrahigh temperatures, you can keep it for a very long time prior to opening the container. This allows large milk companies to buy out the smaller ones and transport your milk all across the country and still get it to your table. The protein is completely denatured

---

LEERY OF LACTOSE?

The good news for the lactose intolerant is that there is much less lactose in aged cheese than in milk. A cup of cow's milk contains about 10 to 12 grams of lactose. An ounce of Swiss, cheddar, or Parmesan (among other hard cheeses) contains less than 1 gram of lactose.

Most of the lactose found in cheese is removed with the whey during the cheese-making process, with the rest being consumed by the culture in the first few weeks of aging.

What's more, when you make your own soft cheeses, they will contain less lactose than store-bought varieties. That's because you will be using live cultures, which eat lactose to produce the acid, and you will not be adding the extra milk solids found in many commercial brands.

As for yogurt, even though it can contain as much lactose as milk, the live active cultures feed on the lactose, leaving little behind. Interestingly, full-fat yogurt has less lactose than low-fat or nonfat. "Dry Curd" Cottage Cheese (page 110) also has almost all of the lactose removed.

and you may as well drink water. You cannot use this type of milk for the wonderful 30-Minute Mozzarella (page 92); it will only make a mushy ricotta. See opposite page for more on UP milk.

ULTRA-HEAT-TREATED (UHT) ASEPTIC MILK

UHT milk, or "long-life" milk, sold in sterilized and hermetically sealed containers, is flash-heated at a temperature between 275 and 302°F (135–150°C). It has a shelf life of more than six months and does not need to be refrigerated until opened. If this is the only milk available to you, you can use it to make soft cheese, but this product comes in a box — need I say more?

WHOLE MILK

Milk still having all of its original ingredients with a butterfat content of 3.5 to 4 percent is called whole milk. Whole milk contains cream.

LOW-FAT MILK

Milk with most of the cream removed, leaving a butterfat content of ½ to 2 percent, is typically called skim or low-fat milk. It is used for making prepared starter and hard, grating cheeses such as Romano and Parmesan. It may also be used as an alternative to whole milk when making fresh and soft cheeses (see chapters 4 and 5) and for a number of other dairy products (see chapter 10).

DRY MILK POWDER

This product is simply dehydrated milk solids; 1⅓ cups of dry milk powder dissolved in 3¾ cups of water makes 1 quart of milk. Dry milk powder does not need to be pasteurized, as the drying process destroys unwanted bacteria. You may use either skim or whole milk powder to make soft cheeses (chapters 4 and 5) and other dairy products (chapter 10).

I have used dry milk powder in the tropics to make fromage blanc with great success. Simply add a packet of starter to the powder, mix it with water, shake, set, and drain, and voilà — island cheese!

BUTTERMILK

Originally, buttermilk was the liquid drained from the churn after butter was made. Little of that is available today. Instead, the buttermilk we buy is made from pasteurized skim milk to which bacterial starter has been added. It is quite simple to use direct-set buttermilk starter to make your own buttermilk (see page 263).

CREAM

There are many types of cream, depending on the butterfat content. If you are buying cream at a store, light cream and half-and-half are ideal for making soft cheeses. Whipping cream and heavy cream are frequently too high in butterfat to set properly. Try to avoid ultrapasteurized cream.

TYPE OF CREAM AND FAT %

Half-and-half	10–18%
Light (or coffee) cream	18–30%
Single cream	20%
Light whipping cream	30–36%
Heavy whipping cream	36–40%

Pasteurization to Ultrapasteurization

First there was raw milk, which is simply the milk as it comes out of the animal. Most cheese makers believe it is the very best choice for making cheese. U.S. law mandates that raw-milk cheeses be aged for at least 60 days to ensure that any pathogens are destroyed.

Next came pasteurization, the process whereby raw milk is heated to 145°F (63°C) for 30 minutes, or 161°F (72°C) for 15 seconds, to destroy pathogenic bacteria, including *Mycobacterium tuberculosis*. As a public health measure, pasteurization has saved countless lives.

As far as cheese making goes, it changes the flavor of the milk slightly and denatures 4 to 7 percent of the whey proteins, which in turn generates a slightly weaker curd. It also makes proteins, vitamins, and milk sugars less available and destroys the enzymes that help people digest milk. For home cheese making, cream-line milk (which has minimum pasteurization and no homogenization) is ideal, but often can be hard to find.

Now there's Higher-Heat Shorter Time (HHST) milk, which is heated to 191°F (88°C) for at least 1 second, destroying most organisms in the milk. There is also an ultrapasteurized (UP) milk, heated to 280°F (138°C), which destroys all organisms. The purpose of some UP treatment is to give the product longer shelf life: UP milk and cream will last 90 days, or longer, unopened. (Once opened, they keep only as long as conventionally pasteurized milk and cream.) It has no real advantage for the consumer but is convenient for the processor, who can buy less milk and transport it farther.

UP milk has a cooked taste, like evaporated milk. UP cream is even worse; it leaves a greasy film on coffee and is difficult to whip. Large processors market UP milk by pandering to people's fears about food safety. Conventionally pasteurized milk and cream are sufficiently treated to deal with any possible pathogens.

Unfortunately, to the dismay of home cheese makers (and consumers who simply like to drink milk), more and more of the milk and cream in grocery dairy cases is ultrapasteurized. Besides giving milk a less-than-desirable taste, the UP process damages the protein structure and destroys the enzymes of the milk so it is utterly useless for a lot of home cheese making. And to make matters worse, you may need to search hard for the small "UP" on the label.

What's a home cheese maker to do? Talk to your grocer and demand an alternative to UP milk. Buy local, and ask the producer what temperature they pasteurize at. Support milk suppliers who are willing to provide a fresh, healthful product. If all else fails, buy a cow, a few sheep, some goats. Drink good milk, and learn to make cheese.

LACTOSE-FREE MILK

If you are able to find a brand of lactose-free milk that is not ultrapasteurized, you can use it to make Whole-Milk Ricotta (page 82) and 30-Minute Mozzarella (page 92), but that's about it. The process of cheese making is based on the bacterial cultures converting the lactose in milk to lactic acid and liquid milk into curds, which eventually become cheese. This conversion also causes the moisture (whey) to be released. Without lactose in milk there is no food to support the bacterial cultures. (See Leery of Lactose?, page 9.)

Rennet

Rennet is a preparation containing the enzyme rennin, which is produced in the stomachs of ruminant mammals. The protease enzyme helps the young digest their mother's milk. Rennin, also known as chymosin, curdles the casein in milk, separating milk into curds (solids) and whey (liquid), thus its use as a cheese-making coagulant. Rennet is used to bring about coagulation while the milk is still sweet.

The amount of rennet used in cheese making varies according to the specific requirements of each cheese, since the curd for each cheese is different. Some types of cheese need a firmer curd than others, and some need a longer time for coagulation.

Rennet comes in liquid, tablet, paste, and powder forms and is available from cheese-making supply houses. Keep liquid and paste rennets in the refrigerator; store rennet tablets and powdered rennet in the freezer. Exposure to light can cause rennet to break down. At temperatures below 50°F and above 130°F (10°C and 54°C), the activity of rennet practically ceases.

Note: "Rennet" is used a little loosely as a term to describe milk coagulants; however, not all coagulants contain rennin.

CHOOSING THE RIGHT RENNET

There is no wrong choice. Because rennet is standardized for its amount of enzymatic activity, all types can be used interchangeably. Liquid and paste are perhaps the easiest to measure, but tablets and powders hold up better under adverse conditions.

Calf rennet is considered the best choice for longer-aged cheeses since some of its residual compounds help complete the breakdown of proteins; some of the more complex proteins in vegetable-based rennet can become bitter in cheese after six months of aging.

Rennet is available in single and double strengths. All recipes in this book are based on single-strength rennet. Follow the instructions on the rennet you are using and adjust the recipes accordingly.

ANIMAL RENNET

Animal rennet is derived from the stomachs of calves, lambs, or kids slaughtered while still on a milk-only diet. It contains two enzymes used in cheese making: chymosin (90%) and pepsin (10%).

VEGETABLE RENNET

Today, most commercial vegetable rennet is obtained from a type of mold (*Mucor miehei*), though there is no mold in the final product. (Liquid vegetable rennet is usually kosher.) Historically, however, vegetable rennet was derived from plants with coagulating properties, which separate the curds from the whey just like the enzymes

MAKING YOUR OWN COAGULANT

If you are really ambitious, you can try making your own coagulant. You will have to see how it sets your milk and adjust the amount used if setting is too fast or too slow. Using standardized coagulants makes your cheese-making process much simpler and more predictable.

When foraging for edible weeds, always check reputable books or online sources for proper plant identification. You can also find some of the plants in natural food stores, fresh or dried: you'll need only about half as much dried as fresh leaves.

To make coagulant using any of the following:

Butterwort leaves (Pinguicula vulgaris)

Knapweed (Centaurea species)

Mallow (Malva sylvestris)

Nettles (Arctium minus)

Our lady's bedstraw (Galium verum)

Stinging nettles (Urtica dioica)

Teasel (Dipsacus sylvestris)

Yarrow (Achillea millefolium)

1. Rinse 2 pounds leaves under cool water and then bring to a boil with 4 cups water and a tablespoon of kosher or sea salt in a large pot, adding more water as needed to cover leaves. Reduce heat, cover, and simmer for 30 minutes.

2. Drain mixture in a cheesecloth-lined sieve, pressing gently on leaves to extract as much liquid as possible. Discard leaves.

3. Let liquid coagulant cool completely before storing, covered, in the refrigerator. Use 1 cup rennet for each gallon of warmed milk.

Note: Wear gloves when working with stinging nettles, as they live up to their name; harvest nettle before it has gone to seed, otherwise it is no longer safe to use in making coagulant.

in animal rennet. In ancient Rome, for instance, cheese makers used an extract of bark from the fig tree (*Ficus carica*). According to legend, at one time in northern Europe, the butterwort plant (cheese flower) was fed to cows just before milking, causing the milk to coagulate three hours later.

THISTLE RENNET

Thistle rennet, a non-GMO flower extract, is available in the United States. Commonly used in Spain and Portugal, the cheeses produced are usually small-scale, artisanal cheeses made with sheep's and goat's milk, such as Ibores.

In the Old Days

In the days before modern laboratory technology produced standardized rennet, most cheese makers made their own rennet on the farm. When they slaughtered a calf or kid, they would clean and salt the stomach and hang it up to dry in a cool place until it was needed. At cheese-making time, they broke off a small piece of the dried stomach and soaked it in cool, fresh water for several hours, then added a bit of the solution to ripened milk to produce a curd.

An even older method called for a calf or kid to be slaughtered at no more than two days old, when the stomach contains milk with a high percentage of colostrum, which is higher in protein, fat, mineral salts, and antibodies. Cheese makers drained the milk, cleaned the stomach, returned the milk to the stomach, tied off the tip, and hung it up to age. Cheese makers also added a tiny amount of finely grated dried cheese to the milk as it was replaced in the stomach, producing a rennet that was both a culture and a coagulant.

JUNKET RENNET

Junket is a soft, puddinglike dessert made with milk and a type of rennet that has one-fifth the strength of regular rennet. Although it is possible to curdle milk with junket tablets, you will have to use five times the amount of junket tablets than regular rennet. It is not recommended for cheese making.

Chlorinated Water

Since chlorine can prevent the action of rennet, you must use nonchlorinated water to dilute rennet. Check with your local water department to find out whether it chlorinates your water supply — something many cities and towns do to kill harmful microorganisms in water. If your tap water is chlorinated, you can either boil and cool it, or leave it uncovered for 24 hours to let the chlorine dissipate. You may also use bottled water or distilled water, available at all grocery stores. If your local water is treated with *chloramine*, it cannot be used for cheese making.

Starter Cultures

Cultures, also known as starters, convert milk sugar (lactose) into lactic acid, bringing about conditions to allow a number of changes to take place. For instance, a starter helps develop the proper level of acidity in milk, which in turn allows the acid to help the rennet coagulate the milk. The acidity aids in expelling the whey from the curds, checks the growth of pathogens, and preserves the final cheese. A starter also contributes to the body, flavor, and aroma of a cheese. When milk has reached the proper level of acidity, it is referred to as *ripened*. All powdered cultures need to be kept in the freezer.

DIRECT-SET STARTER VS. PREPARED STARTER (MOTHER CULTURE)

Traditionally, starters were made from whey, derived from the previous day's cheese making. Over the years, starters were prepared and then propagated from one batch to the next — these were known as *mother cultures* (called "reculturable" starters).

In the late 1980s, a new technology of home starter cultures, known as direct-set starters, was introduced for home cheese making. These are added directly to the milk, which eliminates the lengthy process of making the prepared starter. Direct-set starters save time, are very easy to use, and dramatically reduce the possible contaminants that can enter the milk during the cheese-making process.

You can find direct-set cultures for all types of cheeses as well as for other dairy products at cheese-making supply houses. They are most often in freeze-dried powdered form and can be stored in the freezer (unopened) for up to two years.

MESOPHILIC STARTER VS. THERMOPHILIC STARTER

Although a variety of starter cultures are used in cheese making, they fall into two basic categories: *Mesophilic* (moderate) starters are used to make low-temperature cheeses; *thermophilic* (heat-loving) starters are used to make high-temperature cheeses.

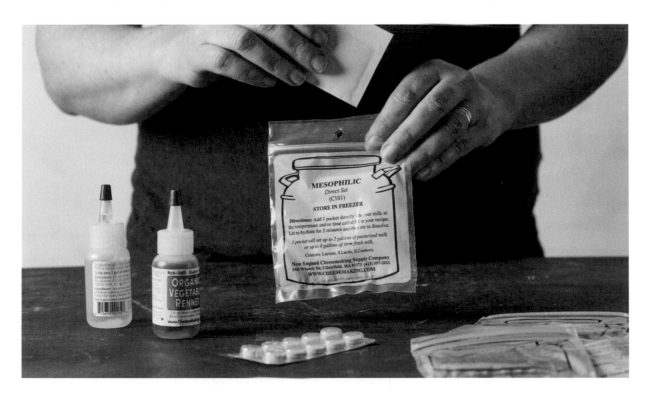

Mesophilic cultures grow best between 86 and 95°F (30–35°C); they are inhibited at 102°F (39°C) and slow at 68°F (20°C). Thermophilic cultures grow best between 102 and 122°F (39–50°C); they are inhibited at 131°F (55°C) and slow or non-existent at 68°F (20°C).

MAKING A PREPARED STARTER. If you prefer to prepare your own mother culture, it may be propagated for one to six months, depending on your attention to cleanliness. To make your first batch of prepared mesophilic or thermophilic starter, add a freeze-dried packet of starter to sterilized, then cooled, skim milk. As time passes, the bacteria reproduce rapidly; after 15 to 24 hours, there are an astronomical number of starter bacteria living in the milk, enough to turn milk protein into a solid white gel. (If you make a mother culture with goat's milk, it will have the consistency of runny buttermilk rather than of gel.)

PROBLEMS WITH STARTER PREPARATION

If you are careful to sterilize everything and monitor the timing and the temperature, probably nothing will go wrong with your starter culture. But . . .

1. If your starter tastes sharply acidic or slightly metallic, it may have overripened. Next time, use a little less starter, or lower the incubation time, or incubate the milk two degrees lower in temperature.

2. If your starter doesn't coagulate:
    - The temperature may have dropped too low during ripening.
    - The inoculating culture lost its viability.
    - The milk contained antibiotics.
    - Residual amounts of bleach on the utensils stopped bacterial action.

3. If you find bubbles in your finished starter culture, discard it. Bubbles may mean the milk and/or your equipment was not properly sterilized. Gas-producing organisms, such as yeasts and coliform bacteria, were present due to faulty preparation.

If you have *any* reason to believe your starter culture is not quite right, *throw it out and begin again with a fresh culture*. It would be heartbreaking to wait six long months for your cheese to age, only to find the wrong bacteria have spoiled it. This is the main reason most home cheese makers use direct-set starter cultures.

1. Sterilize a clean 1-quart canning jar and its lid by placing them in boiling water for 10 minutes.

2. Cool them and fill the jar with fresh skim milk, leaving ½ inch of headspace. Tightly cover the jar with the sterilized lid.

3. Place the jar in a canner (or a big deep pot) with the water level at least ¼ inch above the top. Bring the water to a boil for 30 minutes.

4. Remove the jar and let it cool, away from drafts, to 72°F (22°C) if making a mesophilic starter or 110°F (43°C) if making a thermophilic starter. (Monitor the temperature outside the jar to avoid contaminating the milk.)

5. Inoculate the milk at 72°F for mesophilic and 110°F for thermophilic by adding the contents of one freeze-dried starter culture packet. Add the powder quickly, cover the jar, and swirl to mix.

6. Place the jar in a spot where the milk's temperature can be kept at 72°F for 15–24 hours (mesophilic) or 6–8 hours (thermophilic) to ripen (coagulate).

7. The culture will have the consistency of a relatively thick yogurt. It will separate cleanly from the sides of the jar and its surface will be shiny.

8. Once it has thickened, chill it immediately. You may refrigerate the starter refrigerated for up to 3 days, or freeze it for future use.

TO FREEZE: Place the prepared starter culture in sterilized plastic ice cube trays (standard size). Fill the trays with the starter culture, cover tightly with plastic wrap, and freeze in the coldest part of your freezer. When frozen, transfer the cubes to an airtight container, label with name and date, and put them back in the freezer for up to 30 days. Longer storage will burst some of the bacteria, making them not viable for cheese making.

Each cube of starter culture is a convenient 1-ounce block, which may be used at any time to make cheese or another batch of starter culture. You can also store starter in the freezer in small bottles with lids, which helps prevent contamination.

Note: If you are making a second batch, in step 5 add 2 ounces of fresh starter or frozen cubes rather than the packet of powder and proceed according to directions.

Cheese Coloring

The characteristic color of a finished cheese is often as much a part of its identity as its flavor. For example, we think of goat's-milk cheese as being pure white and some cow's-milk cheddars as yellow. Goat's milk does not contain any carotene, the pigment responsible for the bright hue, while cow's milk has a relatively high carotene content.

Interestingly, the amount of carotenes in the milk fluctuates depending on the season and what the cow is fed. When cows graze on grass, their milk is deeper yellow; when fed hay during the winter, the milk is lighter in color. (When raised on grain, the amount of carotene is significantly reduced, which is another reason to buy milk from pastured cows.) Regardless of the time of year, when the farmer skims the cream with the color-containing butterfat from the top, the resulting cheeses from the lower-fat milk do not have the same rich color.

Orange cheese was originally considered to be of a higher quality than white cheese, and cheese

ANNATTO-TINTED
LEICESTER, *page 154*

GOAT'S-MILK
CHEDDAR, *page 253*

*Cheese comes in a range of colors
depending on the type of milk used
and sometimes the addition of
annatto or other coloring agents.*

FARMHOUSE CHEDDAR,
page 142

makers discovered they could color their cheeses and command a premium price without having to use premium milk. Early on this was accomplished by adding marigold petals, hawthorn buds, saffron, and turmeric. Artificial (synthetic) dyes later became the norm for many mass-market cheeses, though these are now being phased out due to vigorous consumer demand.

Today, most cheese coloring is made from annatto, an extract from the seeds of *Bixa orellana*, a South American shrub. This safe, nontoxic vegetable dye attaches to milk protein (casein) during the cheese-making process and has no effect on the flavor of a cheese. Note when using annatto you will not see the color until the liquid has been drained out of the cheese, so don't use too much.

Cheese coloring comes in liquid form and may be purchased from cheese-making supply houses. Stored in a cool, dark, dry place, it has an indefinite shelf life.

Cheese Salt

Salt enhances the flavor of cheese. Cheese salt is a coarse, noniodized flake salt that is similar to pickling salt. It is usually added to the curds just before they are pressed and, in some cases, rubbed gently on the outside of a cheese after the rind has formed. Cheese salt is also used to make brine solutions in which some cheeses are washed and soaked.

Cheese salt performs many important functions. It draws moisture from the curd, helps drain the whey by causing the curd to shrink, inhibits the growth of lactic bacteria toward the end of the cheese-making process, and acts as a preservative by suppressing the growth of undesirable bacteria.

Cheese salt is available from cheese-making supply houses. Do not use iodized salt; iodine inhibits the growth of starter bacteria and slows the aging process. *Note:* If you plan to freeze a soft cheese for future use, do not add salt before freezing.

Herbs and Spices

Herbs and spices add a variety of flavors and a touch of color to cheeses. I recommend using fresh herbs and spices in fresh cheeses, though dried versions are fine. Use less dried because the flavor is more concentrated. The amount you use depends on your own taste buds, but don't overdo it, because you don't want the herbs to hide the flavors of your cheese.

Use only dried or dehydrated herbs and spices in hard cheeses (using them fresh can introduce bacteria that can contaminate a cheese as it's aging).

When it comes to spices, it is better to buy them whole and then crack, crush, or grind them yourself whenever you need them; whole spices will retain their flavor longer than ground spices. If you opt for preground spices, note the date of purchase on the label and replenish whenever they no longer have a pronounced aroma, usually after six months. Store all spices in a cool place, dark or at least away from direct sunlight.

Popular spices for hard-cheese making are cumin and caraway seeds. For soft-cheese making, try chives, parsley, thyme, dill, oregano, basil, and sage. I also like to roll soft cheeses in cracked peppercorns and herbes de Provence. (See page 73 for other ideas.) Soft cheeses made with fresh herbs

Both soft and hard cheeses can be flavored with herbs and spices.

should sit in the refrigerator for a day or two so the flavors can permeate the cheese. Allow soft cheeses made with (less aromatic) dried herbs to sit a few days longer.

Calcium Chloride

Calcium chloride is a salt solution used to restore balance to the calcium content of milk that has been heat-treated and stored cold. The heat used in pasteurization decreases the amount of calcium in milk and has an adverse effect on its clotting properties. The amount of calcium in milk also decreases over time during cold storage, which means that the renneting action in milk can be significantly slowed.

Adding calcium chloride helps bring this process back into balance and produces a firmer curd. This is especially important when using vegetable or microbial rennet. Calcium chloride has an indefinite shelf life when stored in a cool, dark, dry place.

Note: Due to seasonal variations in milk, differing fat content, and the effects of cold storage and pasteurization, I recommend adding calcium chloride to all recipes except when making pasta filata cheeses, such as mozzarella, because it can prevent them from stretching properly.

Acids

Although rennet is used to aid coagulation of most cheeses, various acids are also used for coagulation in a number of cheeses. Chapter 5 (Soft and Semi-Soft Ripened Cheeses) contains recipes using the following acidifying agents: vinegar; citric acid; tartaric acid; and lemon, lime, and orange juices.

Citric and tartaric acids can be obtained from cheese-making supply houses.

Lipase Powder

Lipase is an enzyme naturally found in goat's milk, often giving goat's-milk cheeses their deliciously pungent flavors. It is sometimes added to cow's-milk cheeses to create a stronger flavor. It is added to many Italian cheeses and occasionally to blue cheese. Lipase powder can be stored in the freezer for up to six months.

Ash

Sometimes called activated charcoal, ash is a food-grade charcoal used on some soft cheeses (particularly goat's-milk cheeses) to neutralize the surface and create a friendly environment for mold growth. Ash is found at drugstores or through cheese-making supply houses.

Mold Powders

Mold powders are added to mold-ripened cheeses to enhance their flavor and aroma. They are available from cheese-making supply houses. The powders will keep up to 12 months stored in the freezer. (*Penicillium roqueforti* has a shelf life of three to six months and cannot be frozen.)

PENICILLIUM CANDIDUM (WHITE)

Generally used as a surface mold on mold-ripened cheeses, such as Camembert and Brie. It gives the cheese its characteristic appearance, and its rapid spread over the surface inhibits the growth of undesirable molds. Its capacity to break down lactic acids allows it to neutralize the acidity of cheese, thereby influencing taste and structure. It also contributes to the ripening process, especially in regard to flavor.

GEOTRICHUM CANDIDUM (WHITE)

Produces white to cream-colored, flat, sometimes almost transparent colonies that may also appear powdery. Together with *P. candidum*, this mold plays a significant role in the ripening process and greatly influences the appearance, structure, and flavor of soft cheeses, such as Camembert and Brie. It is often used in combination with *P. candidum* to prevent slipped skin. In red-smear cheeses, such as Limburger and Muenster, *G. candidum* helps neutralize the surface of the cheese and stimulates the development of *Brevibacterium linens*.

PENICILLIUM ROQUEFORTI (BLUE)

Produces the typical bluish green mottling and restricts undesirable mold growth in blue cheeses. Its enzymes influence the development of a cheese's piquant taste and creamy consistency. It is available as a powder or a liquid.

BREVIBACTERIUM LINENS (RED)

An important component of red-smear cheeses, such as brick and Limburger. It provides the desired color, protects against unwanted mold growth, and is important in flavor formation.

RIND DEVELOPMENT

Proper rind development is important to the success of white mold-ripened cheeses. *Penicillium candidum* is better suited to cow's-milk cheese, where it breaks down proteins and fats and enhances flavor. However, it sometimes grows too quickly on cheese made with goat's or sheep's milk, creating a layer of liquid under the skin and causing the rind to slip off the cheese.

The slower-growing mold of *Geotrichum candidum* helps eliminate "slipskin" in these cheeses. Although it may be used on its own, *G. candidum* is often combined with *P. candidum* to produce a more full-flavored cheese.

EQUIPMENT

or centuries, people have made cheese using indigenous materials, such as clay, wood, and tree bark for molds; straw for cheese baskets and mats; tea towels for draining; and animal stomachs for renneting. Many of these items are still being used today in places where the art of cheese making has been handed down from generation to generation, and it is fascinating to watch delicious cheeses still being produced in this time-tested fashion. And though improvising is fun and possibly less expensive, be aware that using untested materials means sanitation becomes a lot more difficult. Therefore, for the purpose of home cheese making, this chapter covers more modern-day equipment and utensils.

You must take precautions to scrupulously care for and clean all equipment regularly. Stainless steel, enamel, glass, and food-grade polypropylene are used in the kitchen today. They are easy to keep clean and are the most nonreactive with regard to acidity produced during the cheese-making process. Aluminum and unlined cast-iron pots are not used in cheese making because the acid reacts with those metals, corroding them and producing metallic salts, which, when absorbed by the curds, cause unpleasant flavors. In addition, the corrosion of the pots makes sanitation very difficult.

Essential Equipment

Most of these items can be found in your own kitchen. What you don't have on hand you can easily purchase at a department store or from a cheese-making supply house.

ATOMIZER (OPTIONAL)

An atomizer is a bottle with a fine spray nozzle. It delivers a fine mist of mold solution to the surface of mold-ripened cheeses. Too much moisture may result in the growth of undesirable mold.

BOWL (OPTIONAL)

An alternative to heating a pot of milk in a sink full of hot water is to use a 13-quart bowl. Set the pot of milk in the bowl and add hot water to the bowl. In essence, you are making a hot-water-jacketed cheese vat on your table.

BUTTER MUSLIN AND CHEESECLOTH

You need two types of draining material in cheese making, both of which can be purchased from a cheese-making supply house. Butter muslin, which has a tighter weave than cheesecloth, is a must for draining soft cheeses and can be used interchangeably in hard-cheese making.

Cheesecloth is used to drain curds and line molds for hard cheeses. The cheesecloth you may find in a grocery store is not woven tightly enough for cheese making, and it would be a sad story to lose your curds through those large holes.

Professional-quality butter muslin and cheesecloth are strong enough to wash, boil, and use over and over again. Machine washing with an unscented detergent prior to the first use is highly recommended. After using either type, rinse it in cold water to remove any bits of remaining curd, then wash it right away in hot, soapy water. You can add a little bleach to the water before rinsing or boil the cloth in water with a little washing soda to maintain freshness. Have at least one packet of each on hand at all times.

Soft cheeses must always be drained in butter muslin.

CHEESE BOARDS

These are useful as draining platforms and for making a mold sandwich (see page 44) for cheeses such as Brie, Camembert, and Coulommiers. Six-inch-square boards made with well-seasoned (dried) wood work well. Recommended woods include spruce, larch, clear pine, beech, birch, ash, and bamboo; avoid walnut, redwood, teak, and mahogany. Cheese boards of any size are also used for air-drying and aging.

You need two boards to make a mold sandwich. If a recipe calls for more than one mold, you can use larger boards and put the molds together, or you can set up each cheese in its own mold sandwich.

CHEESE MATS

Reed or food-grade plastic mats are used for draining mold-ripened cheeses such as Brie, Camembert, and Coulommiers. They are used in making a mold sandwich (see page 44) and can be put under a hard cheese while aging on a shelf to allow for more air circulation. Cheese mats may be purchased from a cheese-making supply house. You can also use sushi mats or the squares found in the needlework section of a craft store. It is good to have at least one pair of mats on hand.

CHEESE PRESS

Every hard cheese needs to be pressed. A press can be as simple as two boards and a rock. Any model needs to be easy to assemble and easy to clean, and must provide a way to measure the amount of pressure applied to your cheese. Several types of presses and press plans are available from cheese-making supply houses, or you can build one yourself. You may even be able to wheedle one out of a handy woodworking friend or neighbor — maybe in exchange for some homemade cheese.

CHEESE TRIER (OPTIONAL)

This stainless-steel corer is used to take a plug out of your aging cheese for sampling to determine whether it is properly ripened and ready for eating. (A trier allows you to have your cake and eat it, too, because if you simply cut into your cheese the aging would stop.) Triers are available only through a cheese-making supply house.

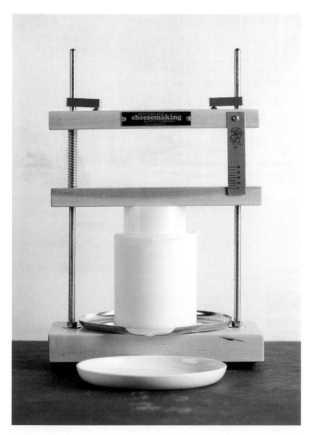

Cheese presses come in several different models.

Curd scoop

Perforated
ladle/skimmer

Cheese mat

Curd knife

Cheesecloth

CHEESE WRAP

Wrapping your cheese in the proper paper creates a controlled environment for optimal ripening. Double-layered papers for mold-ripened and washed-rind cheeses absorb moisture from the surface through the inner layer while the outer layer allows gases to be exchanged and controls further moisture loss. For a very moist cheese, waxed paper can be used until the moisture is reduced. These specialized papers are available from cheese-making supply houses. See Storing Cheese, page 55, for more.

COLANDER

Used for draining the whey from the curds. You can use one of stainless steel or food-grade plastic; just make sure it has plenty of holes on the bottom for good drainage.

CURD KNIFE

Used for cutting the curd, a curd knife has a flat blade with a rounded end rather than a pointed one and is long enough to reach the bottom of the pot without immersing the handle.

DAIRY THERMOMETER

The temperature range of a dairy thermometer needs to be 0 to 220°F (–18 to 104°C). Any thermometer with this range that can be inserted in milk is acceptable. If you need to check its accuracy, dip it in boiling water. If it doesn't read exactly 212°F (100°C), you'll need to make an adjustment before using. A variety of dairy thermometers are available from cheese-making supply houses. The stainless-steel dial-head thermometers respond quickly to temperature changes and most have a bracket to hold them on the side of a pot. (The

Gauging the correct temperature of the milk is a critical part of making good cheese.

temperature range of a candy thermometer is too high for cheese making.)

DRIP TRAY

A tray to put under your mold to catch the whey as it drains. You may use a baking sheet or purchase a drip tray from a cheese-making supply house. The tray also protects the wood surface of a cheese press from moisture while the cheese is being pressed.

HYGROMETER (OPTIONAL)

Used to monitor humidity in your aging area. It is readily obtained at hardware stores and is especially useful for surface-ripened cheeses.

KITCHEN SCALE (OPTIONAL)

A digital scale is useful for measuring ingredients, such as cultures taken from larger culture packages.

LADLE OR SKIMMER

You will need a perforated ladle, also called a skimmer, or a curd scoop to transfer the curds of some soft and mold-ripened cheeses into colanders and molds, to stir cultures and rennet into your milk, to stir in any acidic additives, to mix in your calcium chloride, and to top-stir your milk. You may also pour the rennet through the ladle to distribute it evenly over the surface of the milk. As with all utensils, stainless steel is best. You may find these in a local supermarket or kitchen store, if not already in your kitchen drawer.

LIQUID MEASURING CUP

A glass one is best, because you will be using it for rennet and need to keep it sterile. Plastic may be used, but if scratched, it will be impossible to clean properly.

MEASURING SPOONS

Stainless steel is advised. Besides the standard set, it's worth picking up mini measuring spoons, which

It's important to use the correct mold for the type of cheese you are making.

start with at least $\frac{1}{32}$ teaspoon. Because many starter cultures and other mold powders come in large packs, these mini spoons let you make the most precise micro measurements for the most accurate results. Look for them at cheese-making supply shops.

MOLDS AND FOLLOWERS

Molds come in a variety of shapes and sizes and are used to contain the curds during pressing and the final draining period. A mold determines the shape of your final cheese. The ones obtained from cheese-making supply houses come in stainless steel and food-grade plastic and have drainage holes. (I choose not to use PVC pipe.)

To create a homemade mold, select a food-grade plastic container and punch holes from the inside into the sides and bottom. Make sure you know how large or small the holes need to be for the cheese you will want to make.

Followers are disks designed to fit inside the mold and press the curds evenly by distributing the weight from the press across the surface of the cheese. They are first covered with a small piece of cheesecloth and sit on top of the curds. They are made of stainless steel, food-grade plastic, or wood. If you make your own wooden followers, cut them a hair smaller than the diameter of your mold, as wood expands and contracts with moisture. Use ash or fir; maple tends to mold quickly, and cherry gives off tannins. Don't use pressure-treated wood or any wood treated with chemicals.

NOTEBOOK

Record keeping is an essential part of cheese making. Whether things go wrong or right, you will want to know exactly what you did or did not do. When you

WAXING SUPPLIES FOR AGING CHEESE

CHEESE WAX

You don't need to wax your cheeses; however, waxing does keep hard cheeses a lot moister in the end. A pliable wax creates a protective coating to inhibit bacteria and prevent cheese from drying out during the aging process. These waxes are usually food-grade petroleum-based products; you can also use beeswax.

Cheese wax comes in many colors, most commonly red, black, and natural. You do not want to use paraffin, as it is too brittle and will crack on your cheese, leaving a path for air and thus molds to get in. Cheese wax is reusable and available from a cheese-making supply house.

WAX BRUSH

Make sure you have a natural-bristle brush; silicone brushes will melt right into the wax, leaving you with a nub. Reserve the brush for waxing only, as you will never get it completely clean.

WAX POT

Any type of pot will do for melting wax. If you want to dip your cheese into the wax, a 1-gallon pot will be large enough for most cheeses. If you want to apply wax with a brush, you'll need to melt a 1-pound block, which will fit into a 1-quart saucepan. Don't plan to use these pots for anything else — cleaning them will be nearly impossible.

keep notes, the reasons for variations in your final cheeses will be more understandable. You may want to repeat a "mistake" or a set of conditions that resulted in a particularly delicious cheese! Copy and fill in the Homemade Cheese Record Form on page 355 when making your cheeses. The more you have recorded, the easier it is to get help if you need it later.

POTS

Stainless-steel pots are the best, though glass or unchipped enamel pots may also be used. A pot with a nice thick bottom will hold the heat better than a thin one. You may also choose to heat the ingredients indirectly in a stainless-steel double boiler.

Cleaning and Caring for Your Equipment

Before getting started, thoroughly rinse your equipment and utensils in hot water; sterilizing them is best (see page 30). During cheese making, all utensils coming in contact with milk are first rinsed in cold water, then washed again in hot water. Rinsing them first in cold water prevents any buildup of milkstone, which creates unwanted bacterial accumulation and may contaminate future cheeses.

You will also want to prepare your kitchen before you begin sterilizing your tools. Here are some tips for an extra-clean work space.

- Do not bake bread for at least 36 hours before making cheese to avoid contamination by yeasts. Do not use towels used for bread making for your cheese making.

- Do not have your windows open, especially on windy days.

- Do not let your pets come in and out of the kitchen while you are making cheese, as they are sources of coliform bacteria.

- Do not have any other food on the counter; raw meat can be especially problematic as a potential source of salmonella.

- Do not use dish towels or sponges; put them away and use only paper towels.

Sterilizing Your Equipment

The cleaner your equipment, the more likely you are to have a great cheese-making experience. Sterilizing your equipment before and after use is recommended, but don't be intimidated. Use common sense to keep things clean while making cheese. People have been doing this for thousands of years under all types of conditions.

Sterilizing can be done with the following methods.

- Immerse equipment in a pot of boiling water and heat to at least 185°F (85°C) for 10 minutes. (Alternatively, you can put heatproof equipment

in a 340°F [171°C] oven for 1 hour; or use the sanitizing cycle on your dishwasher without detergent for everything except food-grade plastic.)

- Steam utensils for 10 minutes in a large kettle with about 2 inches of water in the bottom and a tight lid on top. Wooden items, such as cheese boards, may be scrubbed and air-dried. Between uses, mats must be boiled or steamed for at least 10 minutes.

- The most widely used sanitizer for cheese making is Star San. Made from food-grade phosphoric acid, it is flavorless, odorless, and biodegradable. If not rinsed off, the acid residue continues to fight off unwanted bacteria for the entire cheese-making process. When it comes in contact with your milk, it becomes inactive and tasteless. Usage is 2 tablespoons per 5 gallons of water; wipe or spray the mixture on your tools and let air-dry. Keep the mixture in a sealed container for up to four weeks or until it becomes cloudy.

- Sterilize food-grade plastic equipment by dipping it in a solution of 2 tablespoons of household bleach (sodium hypochlorite) per 1 gallon of water. Dampen a clean paper towel in the bleach solution and wipe all work areas.

- You may also use bleach with stainless-steel utensils. Make sure you rinse them thoroughly afterward, because a residue of sodium hypochlorite will interfere with the growth of cheese-making bacteria and may kill the rennet.

- Air-dry all rinsed equipment and store in a clean place. Just before starting a recipe, sterilize all the pieces again and have them handy.

TECHNIQUES

All the techniques presented here are used at one time or another in cheese making. Each type of cheese (e.g., soft, hard, mold-ripened) involves techniques common to all or most of the cheeses within that group. I recommend you read through this chapter to familiarize yourself with the various techniques before jumping in and making cheese. Then, while you follow a recipe, refer to the appropriate sections for a brief review as needed.

The first few pages provide a convenient, at-a-glance summary of making a hard cheese. The steps will vary according to what type of cheese you are making, so not all the steps will be used for all the cheeses.

The remainder of the chapter provides a detailed overview of the techniques specific to each step. The directions are easy to follow, and as you progress, you will get the hang of it relatively quickly.

Hard-Cheese Making at a Glance

1. Warm the milk.

2. Add the starter and calcium chloride.

3. Pour in the rennet.

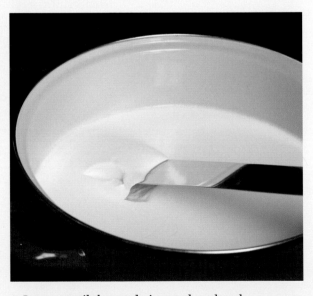

4. Let set until the curd gives a clean break.

5. Cut the curds.

6. Cook the curds.

7. Drain the curds.

8. Mill the curds.

9. Salt the curds.

10. Mold the cheese.

11. Press.

12. Air-dry.

13. Wax (optional).

Final Step: Age.

Ripening

The first step in making cheese is to ripen the milk. In the ripening process, lactose (milk sugar) is converted into lactic acid. The increase of acidity in the milk aids in the expulsion of whey from the curd, helps the rennet coagulate the milk, acts to preserve the final cheese, and assists in flavor development.

To ripen, milk is heated and at the proper temperature cheese starter culture (starter) is added. The starter contains active lactic acid–producing bacteria, which ripen (acidify) the milk, usually over a period of 30 to 60 minutes. To create the perfect environment for acid development, follow the directions for timing exactly.

We control the acidity by planning the ripening time so the level of lactic acid will be just right. If too much acid is produced, the cheese will sour and could leak whey during the aging process. If too little acid is produced, the cheese may have little flavor and can contain numerous gas holes caused by contaminating yeasts or coliform bacteria.

Remember, cheese is alive. The starter must be healthy and active during cheese making. Pay particular attention to preparing and storing your starter (see page 15), and make sure the milk is heated to the correct temperature, at the correct rate, and in the correct manner according to the recipe. During the ripening process, milk is heated in one of two ways: directly or indirectly.

DIRECT HEATING — STOVETOP

Use this method when you see the stovetop icon in the recipe. A medium heat is advised.

With this method you heat your milk directly on the stove. This can heat your milk a little unevenly, therefore is not advised for making hard cheese.

 Direct heating takes place on the stovetop.

 Indirect heating takes place in a water bath.

INDIRECT HEATING — WATER BATH

Use this method when you see the water-bath icon in the recipe. Place the pot of milk into a sink or bowl full of hot water, 10 to 20 degrees warmer than the target temperature of the milk. If the temperature of the milk starts to climb too high, remove the pot from the water. If the temperature falls too low, place the pot back into the water. Add hot or cold water to the sink or bowl as needed to maintain the temperature of the milk.

Adding Starter

A healthy starter is the key to good-quality cheese. Starter converts the lactose in milk to lactic acid to produce controlled ripening. The categories of starter and how to use them are described below.

MESOPHILIC

Mesophilic starter is used in cheeses cooked to moderate temperatures (see page 16 for optimal temperatures). To add a direct-set powdered starter to your milk, tear open the packet, pour the powder onto the milk, let sit for two minutes for the powder to rehydrate, then stir to combine. When using a prepared mother culture, 4 ounces prepared = 1 packet direct-set. Once mixed, do not stir the milk for the remainder of the ripening period; excessive aeration reduces the rate of acid production, which may extend the ripening time.

THERMOPHILIC

This heat-loving starter (see page 16 for optimal temperatures) is used in cheeses cooked to higher temperatures. Follow the directions for mesophilic starter.

PREPARED

To use prepared starter, make your own mother culture, store it, and then propagate as needed (see pages 15–17). Add the required amount of prepared starter to the milk and stir to distribute evenly. Once mixed, do not stir for the remainder of the ripening period; excessive aeration reduces the rate of acid production, which may extend the ripening time.

Mixing in Additives

You may find it necessary or desirable to add various ingredients to enhance flavor, provide color, or aid coagulation. The timing of when to mix in these additives depends on the recipe.

It is a good idea for beginning cheese makers to sterilize the water used to dilute various additives. This eliminates one variable and rules out contamination at this step if you have a problem with your final cheese. To sterilize water, bring it to a boil, cool it, and store it in a sterilized bottle in the refrigerator to use as needed. Below are the various types of additives and how to use them.

Calcium chloride and rennet must be diluted before being added to milk.

CALCIUM CHLORIDE

Calcium chloride helps calcium bond with the protein to get a good curd rather than a soupy mess. It is currently recommended for use in all of your cheese making except the pasta filata family of pulled-curd cheeses such as mozzarella and provolone, which need a partial loss of calcium to stretch. Use ¼ teaspoon calcium chloride per gallon of milk, diluted in ¼ cup of nonchlorinated water. Add the mixture when you begin heating your milk.

COLORING

Cheese coloring can destroy the coagulating ability of rennet, so it is added to ripened milk before the rennet. Remember: you are adding the coloring to milk when there is still a lot of water in it, therefore the color will not be apparent until you drain the curds.

Dilute coloring in 20 times its own volume of cool water (¼ cup of water is usually sufficient for the amount of coloring you'll use) and thoroughly mix it into the milk. If it is not mixed adequately prior to adding the rennet, it may cause streaking in the finished cheese.

If you plan to use the same bowl or cup to dilute the rennet, make sure you wash it thoroughly. The annatto in the coloring, if concentrated, may weaken the action of the rennet.

LIPASE

Lipase enzyme produces additional acid in milk, so it may be necessary to adjust the amount of rennet used, depending on the firmness of your curds.

Dissolve lipase in ¼ cup of cool water and let it sit for 20 minutes before adding the mixture to the milk.

MOLD POWDER

These are added to mold- or surface-ripened cheeses to produce the characteristic flavors and textures of the cheese.

Brevibacterium linens (*Bacteria linens*) is a bacteria added to the milk at the same time you add your culture. In some cheeses, it is smeared by hand on the surface of the cheese after brining. To do this, gently wash the cheese each day with a light brine (page 46), as per the recipe. Wearing sterilized gloves, wet the palm of one hand with the brine and dampen all surfaces of the cheese.

Geotrichum candidum is often added to the milk in conjunction with *Penicillium candidum* at the same time you add your culture.

Penicillium roqueforti, a blue mold powder (also available in liquid form), is often added to the milk at the same time you add your culture.

Renneting

In this step, rennet is added to the ripened milk. Temperature is important, so follow the recipe carefully. The milk is left to set until it has coagulated. The enzyme rennin (or, in the case of vegetable rennet, a microbial enzyme) causes the protein portion of the milk to precipitate out of solution, becoming a solid white custardlike mass called curd. Trapped within this mass of curd are the butterfat and whey. The whey contains water, milk sugar, albuminous protein, and minerals, and it may be saved to make other recipes (see page 45).

Measure rennet carefully. Too little will not set the milk properly; too much will result in a rubbery and possibly bitter-tasting cheese. The perfect curd is one that produces the highest yield of cheese from any given amount of milk.

DILUTING RENNET

Always dilute rennet (liquid, powder, or tablet) in cool, nonchlorinated potable water (¼ cup is sufficient for a 1- or 2-gallon recipe) before adding it to the ripened milk. If using powdered rennet, let sit for 30 minutes in the water, stirring occasionally, to completely dissolve it before use. If using a rennet tablet, crush it with the back of a spoon and let sit for 10 to 30 minutes in the water to completely dissolve before use. If rennet is not diluted, it will be unevenly distributed in the milk, which could produce a faulty curd.

DETERMINING COAGULATION AND SETTING TIME

If you notice how long the rennet takes to start coagulating the milk, you can estimate the setting time to be two and a half times that long. However, the higher the moisture content in the final cheese, the longer it will take.

To determine the flocculation point, which is when the milk has begun to coagulate, put a drop of the milk in a small glass container of water. If the milk simply disperses, it has not started to coagulate yet. If you see small globules floating in the water, coagulation has begun.

ADDING RENNET

Pour the diluted rennet solution into the milk through a perforated ladle to help disperse it evenly. Mix gently with an up-and-down motion for 30 seconds, making sure you reach the bottom of the pot. Cover the pot and let the milk set for the amount of time specified in the recipe.

Do not stir the milk once it has started to coagulate; doing so will cause severe loss of butterfat. It is also very important not to disturb the pot while the curd is setting because it may break the bonds forming between protein molecules and they will not rejoin, thus producing a weak curd.

TOP-STIRRING

If using raw milk, top-stir it for another 30 seconds when adding rennet. This mixes any butterfat that has risen to the surface back into the body of the milk. To top-stir, simply stir the top ¼ inch of the milk with the bottom of a slotted spoon or perforated ladle.

Top-stirring takes place just below the surface of the milk.

Cutting Curds

The curd is carefully cut into small, uniform cubes to increase the surface area, which allows the whey to continue draining. Uneven or crushed curds have much more surface area, resulting in more rapid and uneven drainage and increased loss of butterfat and solids. It's important to cut the curd at the right moment. If you cut it too soon, it will be soft and unworkable; if you wait too long, it will become too firm. If it is too firm, the knife will crush the curd rather than cut it cleanly.

The curd is ready for cutting when it gives a clean break. This can be determined by placing the tip of a curd knife (or a clean finger) into the curd at a 45-degree angle. If the curd separates in a clean line, you have a clean break and the curds are ready for cutting. If the curd does not show a clean break, wait five minutes longer and test it again.

Immediately after cutting, the whey will be a whitish hue because of the small amount of butterfat that escapes during cutting. Shortly after, the butterfat will seem to disappear and the whey will take on a translucent greenish hue.

Note: A good rule for cutting the curd is to time how long it takes for the curd to set and multiply by three. For example, if the curd first sets in 12 minutes, the cutting time is after 36 minutes.

Cutting Hard-Cheese Curds

Different varieties of cheese get cut either into larger or smaller cubes depending on the final moisture level. Each recipe specifies the desired size of the curd and resting time.

1. For the first cut, place your curd knife in the pot all the way to the bottom; gently draw the knife through the curd in a straight line, making sure your knife is always touching the bottom of the pot. Continue cutting vertically in this manner all the way across the curd.

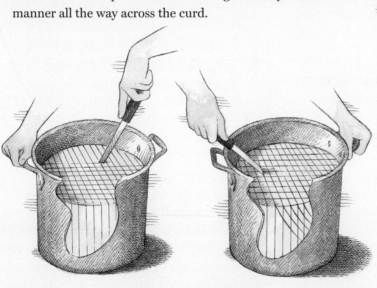

2. Now cut in the same manner horizontally all across the pot. This creates a checkerboard of columns.

3. Wait the specified time to let some whey release and allow the curds to heal. Then, with the knife at a 45-degree angle and using the previously cut lines as your gauge, cut across those lines at a 45-degree angle, going all the way across the pot.

4. Turn the pot 90 degrees and repeat.

Note: A large stainless-steel whisk can also be used in steps 3 and 4.

Newly formed curds must be handled with great care.

LADLING AND SLICING SOFT AND MOLD-RIPENED CHEESE

Gently ladle thin slices of your curd into a mold or a butter muslin–lined colander. The less you break up the curds at this point, the creamier your final cheese will be.

CUTTING HARD CHEESE INTO CUBES

See Cutting Hard-Cheese Curds on page 41. Your aim is to cut the entire curd into specific-sized cubes as called for in the recipe. You want to have the curds as uniform as possible, but don't be alarmed if there is some variation in the size of your curd pieces.

When all the curd has been cut, let them sit for 3 to 5 minutes to firm up. Then very gingerly stir the cubes, turning them over from bottom to top. Cut any pieces (referred to as whales) that are still too big. Some recipes may call for the second two cuts (steps 3 and 4 on page 41) to be done with a stainless-steel wire whisk. Be careful not to over-stir — butterfat is trapped in the curds, and stirring too vigorously will result in a loss of butterfat and produce a cheese of poor consistency.

Cooking Curds

After the curds have been cut into uniform cubes, you will notice the cubes floating in more liquid.

Each cube is slowly losing whey and shrinking in size. At this point, the cubes are soft and jellylike, and they must be handled very gently to avoid damaging the curd and losing butterfat. The cubes are now heated indirectly to expel more whey, to firm and dry the curds, and to increase acidity. Too much acid may result in a sour and/or bitter cheese, with a moist, soft texture. Too little acid can produce a cheese with little flavor.

As with ripening, the temperature of the curds is controlled by the temperature of the water surrounding the pot. The curds are gently stirred every three minutes to keep them from matting. For many cheeses, it is important to increase the temperature of the curds by no more than two degrees every five minutes. If the curds are warmed too quickly, they will develop a thin skin, trapping the whey inside and preventing adequate drainage, which may leave the final cheese with excess moisture. As the temperature slowly increases, the cubes shrink and firm up as they expel more whey. The amount of whey in the pot will noticeably increase.

Hard-cheese curds are ladled into a colander lined with cloth for draining.

Draining

After the curds are cooked, they are drained to remove more whey. The draining process varies depending on the type of cheese being made.

Note: The whey from hard cheese may be saved for further cheese making, but it must be used within three hours of collecting it or it becomes too acidic to use.

SOFT CHEESE

Soft-cheese curds are typically placed in butter muslin (not cheesecloth, which is too loosely woven) and hung to drain.

Line a colander with butter muslin and gently ladle in the curds. Tie the corners of the muslin into a knot and hang the bag over the sink or a bowl. A bungee cord hung from a hook over the sink makes a great draining tool.

Note: Draining in a colander lined with butter muslin is acceptable if the colander has enough holes for drainage. Otherwise it may slow down the draining, resulting in a more acidic cheese.

HARD CHEESE

Hard-cheese curds drain for a variety of times, depending on the recipe. Line a colander with cheesecloth or butter muslin and carefully ladle the curds into the colander.

SURFACE- AND MOLD-RIPENED CHEESE

For many of these cheeses, the curds are ladled directly into the mold to drain under their own weight, rather than by being pressed (see page 48). Drainage takes place through the holes in the mold and the mats on which the cheese sits.

A "mold sandwich" is often used in draining mold-ripened cheeses. The mold is placed on a cheese mat that rests on a board placed in a sink or a baking pan. The mold is filled to the top with curd and drainage begins immediately. It is often necessary to let the curds settle, then add more as the draining takes place. Follow the directions in the recipe carefully. After the mold is full, a second cheese mat is placed on top of it and a second wooden board is placed on top of the mat.

After a period of draining, the entire assemblage is flipped over. The cheese mat now on the top is gently peeled away, which allows the cheese to fall gently to the bottom of the mold. The cheese mat is then replaced and the process is repeated at regular intervals.

Milling Hard Cheese

After draining, some hard-cheese curds are broken into small pieces without squeezing out any moisture. This requires great care; otherwise, a serious loss of butterfat will occur.

With your hands, break the curds into small pieces ranging in size from a thumbnail to a walnut, depending on the recipe's directions.

MAKING A MOLD SANDWICH

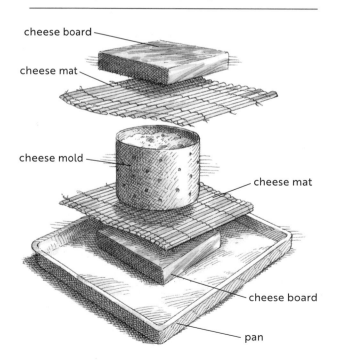

cheese board

cheese mat

cheese mold

cheese mat

cheese board

pan

Milling hard cheese

Salting

Cheese salt enhances the flavor of the cheese and acts as a preservative by suppressing the growth of undesirable bacteria. It helps inhibit the growth of lactic bacteria toward the end of the cheese-making process. The salt also draws moisture from the curds, causing them to shrink and drain more whey. Always use proper cheese salt, not one with additives such as iodine or dextrose.

When making aged cheeses, you may be surprised to learn that the more salt you add to your cheese, the less salty it may taste once it is aged.

WAYS TO USE WHEY

There are two types of whey after draining, depending on the type of cheese being made.

SWEET WHEY (usual pH 6.7) is drained from hard cheeses before the bacteria changed most of the lactose into lactic acid. Most of these cheeses drain the majority of their whey within an hour. This whey needs to be used within 3 hours of draining or it will become too acidic as the bacteria continue turning the lactose into lactic acid. The sweeter the whey, the richer the cheese.

You can use whey to make other cheeses (see Whey Cheeses, page 254). Substitute it for all or part of the water in soup stock or for baking (see Italian Feather Bread with Whey, page 305). Use it to make mashed potatoes, pancakes, and smoothies. Feed it to plants in the garden — use it undiluted for tomatoes and mixed half and half with water for acid-loving plants. It is also a great to feed to pigs and chickens.

ACID WHEY (usual pH below 5.1) is drained from cheeses set with an acid, such as vinegar, tartaric acid, citric acid, or lemon juice, or any soft cheese or other dairy product that takes longer than an hour to drain, such as fromage blanc, sour cream, and mascarpone. This whey usually has a higher calcium content. You can use it sparingly to feed acid-loving plants, but it is too acidic for most culinary purposes.

This is because salt slows down the aging process, which naturally produces its own salts.

There are different ways to salt cheese, depending on the type being made.

DIRECT SALTING

In this case, salt is added directly to the curd after draining. For soft cheeses, remove the cheese from the muslin bag, salt to taste, and mix well. For hard cheeses, sprinkle cheese salt over the milled curds according to the recipe's directions, and gently but thoroughly mix it into the curds. For mold-ripened cheeses, rub salt on the outside of the cheese prior to drying and aging.

BRINING

Brining is used mainly for cheeses that have a short aging time. Some cheeses, such as feta and Gouda, are not salted before pressing; rather, they are put into a brine bath after pressing. Many other cheeses require a period of brining after pressing.

MAKING BRINE FOR CHEESES

Because brine loses some of its salt to the cheese and may become contaminated with unwanted bacteria over time, it's important to use and store it properly as noted.

SATURATED BRINE

- 1 gallon nonchlorinated water
- 2¼ pounds cheese salt
- 1 tablespoon calcium chloride
- 1 teaspoon white vinegar

Combine the water, salt, calcium chloride, and vinegar in a large nonreactive pan set over medium heat. Bring to a boil and stir until the salt dissolves. Cool and store in a glass jar in the refrigerator for up to 2 years.

After using a brine bath, bring it to a boil and add additional salt, stirring until the salt no longer dissolves in the water. If it gets cloudy, filter it through butter muslin and boil for 10 minutes and cool before using.

LIGHT BRINE

A light brine is used to wipe down the surface of aging cheeses if any unwanted mold appears.

- 1 cup nonchlorinated water
- 1 tablespoon cheese salt

Combine the water and salt in a small jar, stirring to dissolve the salt.

A brine bath may range from lightly salted water to a saturated salt solution (in which the water is so saturated with salt that some salt precipitates out of the solution). The amount of salt needed for a brine solution depends on the type of cheese being made.

Cheese should always be cooled down before brining, as a warm cheese in brine will become too salty. Both the cheese and the brine should be between 50 and 55°F (10–13°C) and they should have the same pH level.

Molding Hard Cheeses

Select the mold according to your recipe and line it with cheesecloth. (Remember the mold determines the shape of the cheese. If you use one too large for the amount of curd, your cheese could turn into a pancake.)

Place the curds into the mold. For cheeses being pressed lightly or not at all, pack the mold lightly, so the curds will not lose butterfat. For cheeses subjected to high pressure (most hard cheeses), pack the curds firmly. In most cases, the curds must not be pressed before their temperature has fallen to 70°F (21°C) or lower or much of the butterfat will be lost, ruining the flavor and texture of the cheese.

To test hard-cheese curds for the proper consistency for molding, scoop up a small handful and squeeze them gently They should hold together in a single lump.

Pressing Hard Cheese

The cheese is pressed for varying lengths of time to compress the curd and expel additional whey. The amount of pressure determines the final texture of your cheese. It is important to apply pressure to cheese gradually. Heavy pressure at the beginning will cause a lot of butterfat to be expelled, forming a hard coat and preventing the cheese from draining properly.

Line a mold with dampened cheesecloth. Place the mold on a drip tray, which will allow the whey to drain into a sink or small container. Ladle the curds into the mold. Lay a piece of dampened cheesecloth on top of the curds and cover it with a follower that fits the mold. Once the follower is in place, pull the cheesecloth snug to eliminate any bunching of the cloth.

Generally, you will apply light pressure to the cheese for the first 10 to 15 minutes. Then remove the mold from the press, remove the cheese from the mold, gently peel the cheese-cloth from the cheese, and flip the cheese. Wrap the cheese in the same cloth to prevent the cloth from sticking to the cheese, a process called re-dressing. Place the cheese back into the mold, put on the cloth and follower, and continue pressing. Turning the cheese results in even pressing. The pressure is increased to the recipe's directions.

The cheese is usually flipped several more times and pressed at increasing amounts of pressure, until it is finally left in the press at a maximum pressure usually for at least 12 hours.

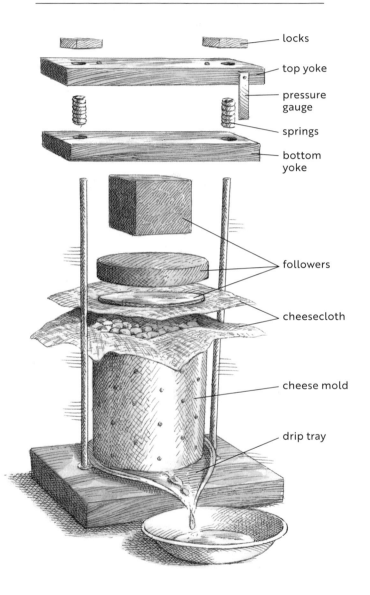

PARTS OF A CHEESE PRESS

- locks
- top yoke
- pressure gauge
- springs
- bottom yoke
- followers
- cheesecloth
- cheese mold
- drip tray

Air-Drying

Air-drying prepares hard cheeses for aging. Remove the pressed cheese from the mold and gently peel away the cheesecloth. Place the cheese on a cheese board or cheese mat and, turning it several times daily, air-dry at room temperature for several days, or until it is dry to the touch on all surfaces.

After the cheese has dried properly, you may choose to oil the surface during the aging process.

This keeps the cheese from drying out and protects against mold growth. Oiling is often used on cheeses 2 pounds and over because it allows a beautiful rind to develop. Waxing (see page 50) prevents a rind from developing; it protects smaller cheeses from drying out, thereby creating a moister cheese.

If any unwanted mold develops on the surface of the cheese, dip a piece of cheesecloth into a light brine, ring it out so it is merely damp, and rub the surface of the cheese to remove the mold.

BANDAGING HARD CHEESE

After air-drying, hard cheeses are aged. You can try bandaging your cheeses for a deliciously pronounced flavor. This method can make a pretty fabulous cheddar. It is perfectly normal for mold to develop on bandaged cheeses; the mold will protect the inside of your cheese from drying out while allowing it to breathe and develop optimal flavor.

Cut a piece of butter muslin as wide as the depth of the cheese and 1½ times its circumference in length. Cut four circular pieces to act as caps for the top and bottom, making them larger than the cheese so they will fold over the sides. Rub a thin coat of lard or solid vegetable shortening on the cheese and place two caps at each end. Wrap the bandage around the cheese, sticking it down as you go.

This larding technique is a traditional English method that produces a drier, flakier cheddar than the buttery-textured cheddars

more commonly produced in the United States. Store the cheese at the temperature indicated in the recipe and keep it bandaged until it is ready to be cut and served. The lard and mold on the outside of the bandage will come off when you remove the bandage.

Waxing Hard Cheese

Once a cheese is dry, it will automatically develop a nice natural rind without waxing. However, you can wax a cheese to keep it from drying out too much and to retard the growth of mold during the aging process. If you make small cheeses, you may want to wax them to prevent excess drying and a hard texture.

Cool the cheese in the refrigerator for several hours prior to waxing, so that the wax will adhere better. Melt the cheese wax in a dedicated wax pot on a stovetop vented with a hood fan. Wax vapors are highly flammable, so use caution when melting it and do not put a cover on the pot. Heat the wax to 240°F (116°C) to kill any surface bacteria and evaporate excess moisture on the cheese.

Apply the wax with a natural-bristle brush, working on one surface of the cheese at a time. Allow it to cool for several minutes before turning it over and waxing the other surfaces. Cheese wax dries very quickly. It takes at least two thin coats of wax to protect the cheese. Subsequent coats of wax may be applied as soon as the first coat dries.

Alternatively, dip your cheese into the wax pot one side at a time with a quick in-and-out motion so that you do not melt the previous layer. Use extreme caution in the dipping process — the wax is very slippery. Allow the wax to dry between coats.

For accurate labeling, write the name of the cheese and the date it was made on a piece of paper, and wax it onto the cheese when you apply the last coat. If mold growth appears on the surface of the cheese, cut it off before eating. It will not hurt the cheese. (If you plan to smoke the cheese, smoke it first, then wax it. See Smoking, page 53.)

Reusing Cheese Wax

Cheese wax can be used over and over again. Peel the wax off your finished cheese, melt it, and strain it through butter muslin before storing it in the pot at room temperature.

Aging

During this stage, cheese ripens and develops its flavor and character. Aging can take anywhere from just a few days to six years. Store cheese in a temperature- and humidity-controlled environment to promote proper maturation. The temperature at which you age your cheese is important in allowing the good bacteria introduced by the starter to grow and produce enough acidity to preserve your final cheese and to create its characteristic flavors.

Most hard cheeses prefer an aging temperature between 46 and 60°F (8–16°C) and a relative humidity of 75 to 95 percent. The constant exchange of ripening gases from the cheese (including carbon dioxide and ammonia) and oxygen in the air is critical for flavor development.

If you do not have a suitable location for aging, a spot where the room temperature will not exceed 68°F (20°C) is preferable to the household refrigerator. If the aging temperature is too cold, the starter culture won't produce the proper acid development for safe and flavorful cheese. If the aging temperature is too warm, the cheese may develop a sharper, more pungent flavor than is pleasing and the texture may begin to deteriorate. Also, the cheese may develop undesirable mold growth and will require more frequent checks.

Mold may also develop if the aging room is too damp. If the aging room is not damp enough, however, the cheese may shrink and possibly crack. If small cracks appear, simply smear them with a little softened butter.

The precise details for aging vary according to the type of cheese being made; closely follow the recipe directions for aging.

SOFT CHEESE

Soft "bag" cheeses are typically not aged, but you may want to do some experimenting once in a while and lightly press and age one or two for a short time. Some recipes, such as those for Gervais and Bondon (see pages 74 and 75), include more detailed information for pressing and aging.

HARD CHEESE

Store hard cheese on a clean cheese board in a spot where the temperature will remain at a constant 55°F (13°C) and have a relative humidity of 65 to 85 percent. Many home basements will satisfy this requirement.

You might consider purchasing a secondhand wine cooler or a small dorm-size fridge. Place a bowl of water in the bottom of the refrigerator and keep the setting at 55°F, and you will have an ideal cheese-storage chamber. If you use a normal-size refrigerator, you can maintain the desired humidity by putting the cheese into a lidded container about three times the size of the cheese. A cake dome works well. After a while, moisture may condense on the inside cover of the container. If it is enough to drip or pool, wipe the excess moisture out.

Turn over the cheese each day for the first few weeks, and several times a week thereafter. This prevents moisture from accumulating on the bottom of the cheese, causing rot.

The longer a hard cheese ages, the stronger its flavor becomes. (For retail sales, a minimum of 60 days is required for hard cheeses prior to selling them.) Some of the hard grating cheeses, such as Parmesan, may be aged for years to develop a very sharp flavor. If you are beginning this art, you might be tempted to taste your cheese sooner, but don't cut into it too early. Instead, use a cheese trier and your cheese will continue to age after you put the plug back in.

Using a Cheese Trier. After a month or more of aging, a trier is sometimes used to sample a hard cheese to determine whether it has aged sufficiently. Judges in cheese competitions use triers to ascertain information on flavor, aroma, acidity, butterfat and moisture content, texture, and body without having to cut into the cheese.

Push the trier into the side of the cheese (as you do this, note the resistance of the curd). Turn the trier one or two complete turns and carefully withdraw the plug of cheese. Remove a small piece of the plug and roll it between your fingers to warm it. Smell and taste the warmed piece of cheese for aroma and flavor (an "off" flavor is easily detected at this stage). Replace the plug carefully and seal the gap around the plug by smearing a piece of the warmed cheese from the plug with your finger. Continue aging, if necessary.

SEMI-SOFT AND MOLD-RIPENED CHEESE

These cheeses are aged in cool, moist temperatures to allow the prolific growth of mold on both the surface and the interior of the cheese. An average starting temperature of 55°F (13°C) is used for the first week or two until the mold starts growing. Further aging takes place around 45°F (7°C) with a relative humidity of 85 to 95 percent. For a detailed description of the aging technique for semi-soft and mold-ripened cheeses, see page 117.

A cheese trier allows you to taste your cheese as it ages without having to cut into it.

SMOKING

Smoking your cheese is another way to add flavor. The smoke evaporates moisture, bringing butterfat to the surface of the cheese; when combined with the smoke, which contains antimicrobial substances, the butterfat has a preservative effect, provided the cheese is kept dry. In addition to a wonderful smoky flavor, this procedure also imparts an attractive light brown color to the surface of the cheese. (*Note:* If waxing a cheese, wait until after you smoke it.)

The key to smoking is low heat. Smoking should not heat the cheese, which could cause it to lose butterfat. Hang the cheese in a cold, smoke-filled enclosure at a temperature no higher than 40°F (4°C). Oak and apple wood shavings are two appropriate woods for this purpose. Smoke the cheese for four to six hours; an hour or two longer for milder cheeses.

If you do not have a smoker, use this simple smoking method: Put some dampened hardwood sawdust into a metal pan and place it on a few warm coals that have been allowed to burn down on the bottom of your outdoor grill. Set the cheese on a rack well above the smoking material. Cover the grill and allow the smoke to escape through the vent in the cover. Keep a close eye on your cheese to make sure it doesn't get too warm (above 40°F). Smoke the cheese until the exterior is golden brown.

Temperature

As is probably clear by now, temperature plays an important role in the cheese-making process. Although the temperature indicated at each step in a particular recipe is meant to provide the conditions necessary for the successful production of that cheese, experience will teach you there is room for experimentation.

A recipe presents general guidelines to follow when learning to make cheese. Once you master the techniques, you may experiment with varying the temperatures up or down by as much as five degrees to produce subtle changes in the final cheese, turning your hobby into an art form.

Doubling Recipes

When you make one pound of cheese, you use the same techniques and ingredients you would use if you were making one thousand pounds of cheese. Once you become comfortable with a recipe and

TEMPERATURE TROUBLESHOOTING

These are some of the issues that can arise when the temperature isn't correct.

Procedure	If Temperature Is Too High	If Temperature Is Too Low
Starter preparation	Kills active bacteria or makes for very slow acid development	Inhibits bacterial growth
Ripening	Produces a milk that is overly acidic; may kill bacteria	May not be warm enough to allow starter to grow
Renneting	Interferes with starter growth; may decrease rennet activity; at temperatures higher than 130°F (54°C), rennet activity virtually ceases.	May result in a curd that takes too long to set and is too soft to cut
Cooking	Scalds the curds, creating a skin that prevents proper drainage	Interferes with milk's ability to produce the proper amount of acidity
Aging	Encourages undesirable mold growth and perhaps too rapid an acid development	Prevents the proper amount of acidity from building up. As proper acid development is important to the safe preservation of your cheese, this is an important consideration.

know how long ripening and coagulation take, you may increase the recipe. The preparation time is the same; all you need are larger pots and molds.

The proportion of ingredients to add to the milk depends, in part, on the quality of the milk. A good rule is to increase the ingredients proportionately; however, you will have to check the speed of coagulation and adjust the rennet in future trials so you use the amount that results in coagulation at the normal time for the recipe. This often means you will need less rennet.

Acidity Testing

The times and temperatures indicated in the recipes throughout this book are designed to provide the proper levels of acidity required for successful home cheese making.

There may come a time when you find yourself intrigued by the fascinating science behind cheese making and would like to delve into the specifics of acidity. You will want to test for acidity at various stages during cheese making. Determining acidity ensures optimum conditions are met to obtain uniform results.

Acidity is determined by measuring either pH value (true acidity) or titrated acidity. The pH of milk is determined by its content of hydrogen ions and is measured by simply comparing pH strips to a standardized color chart. A value of 7.0 indicates neutral pH. Values above 7.0 indicate an alkaline condition; values below 7.0 indicate an acid condition.

Testing for titrated acidity is a bit more involved. Basically, a few drops of a color indicator,

called phenolphthalein, which is colorless in acid, are added to the milk. As long as the milk is in an acid state, the indicator will remain colorless. The moment the liquid becomes alkaline, the color changes to pink, indicating the instant when the acid was neutralized. Titrated acidity is therefore the amount of an alkaline solution used to neutralize an acidic liquid. Fresh milk has a titrated acidity of 16 to 18. Acid-testing kits, pH meters, and pH strips are available from cheese-making supply houses.

Storing Cheese

Because cheese is a living, breathing organism, it continues to ripen from the day it is made to the moment you eat it. Temperature and humidity during storage are very important because they affect taste and texture, as is allowing the cheese to continue to "breathe" while being stored. For that reason, tightly wrapping the cheese in plastic wrap is often not the best idea.

Tightly wrapping even a finished cheese in plastic wrap is not a great idea. Instead, to keep your cheese in optimal condition, use cheese wrap (see page 27), wax paper, or aluminum foil. Some cheese makers advise perforating the wax paper or foil, but you will need to make sure the cheese isn't drying out too much.

There are also papers specifically designed for mold-ripened cheese: The inner layer, which pulls moisture away from the surface of the cheese, is coated with paraffin to keep the *candidum* mold from growing onto the paper, which would result in tearing of the cheese surface upon opening. The outer layer allows gases to be exchanged while controlling the moisture loss, essentially allowing the molds to remain active while not becoming excessive.

For washed-rind and other hard cheeses, the two-ply papers are not coated with paraffin. The inner layer wicks moisture away from the surface of the cheese to inhibit surface mold growth and the outer layer allows gases to be exchanged while minimizing moisture loss.

As for where to store the cheese, that depends. In an earlier time, before insulation and central heating, many homes had a buttery, or cool pantry, on the north side of the house, where cheese, butter, milk, and even meat were stored during much of the year. Today, lacking that, we have to turn to our refrigerator. Most sources advise refrigerating cheese at temperatures of 38 to 42°F (3–6°C). If your refrigerator has a little cubby in the door that is labeled CHEESE, don't use it. Instead, store cheese in one of the vegetable bins in the bottom of the refrigerator, where it will be out of the airflow. Check the temperature in the bin to be sure it isn't too cold.

In addition, remember that the firmer the cheese, the longer it will stay fresh when treated properly. Although hard cheeses, such as aged cheddar and Parmesan, can be kept (wrapped) in your refrigerator for many months, most soft cheeses, especially after they are cut, will last only two weeks.

Cheese needs the moisture it has, not more or less, so monitor the cheeses you are storing. Check the cheese frequently to see how it is doing; if it gets moldy, trim off the mold and rewrap the cheese in clean cheese wrap, wax paper, or foil. If a semi-hard or a hard cheese becomes dry or cracked on the surface, wrap it for one to two hours in a slightly damp towel dipped in light brine.

KEEP GOOD NOTES

It's important to keep records on every batch of cheese, noting amounts of ingredients, adjustments to the recipes, temperatures, and timing of each step. Good notes allow you to repeat the results if you love a cheese or try again if you want to improve it. See page 355 for an example.

Properly wrapping cheese for storage preserves its flavor and prevents it from drying out or becoming too moist.

DATE Nov. 12

CHEESE Muenster

circle all that apply

CHEESE Stilton

circle all that apply

DATE Jan. 10

circle all that apply

CHEESE

DATE

A CHEESE MAKER'S STORY

Jeff and Jenna Fenwick

BACK FORTY ARTISAN CHEESE

Mississippi Station, Ontario

Jeff Fenwick

WHEN JEFF AND JENNA FENWICK decided to leave urban life for a more rural environment, they started by exploring their money-making options. His background was in sales and marketing for a university, she was a textile designer with her own business. Having toyed with the idea of opening a pub or restaurant, it wasn't too much of a leap for them to snatch up an existing cheese-making enterprise near where Jenna's parents lived and where Jeff had attended college. But rather than being able to slowly ease into cheese making, Jeff had to keep up with the many orders for the cheese from the get-go, with just a few weekends of training from the previous owner (and some prior beer-making experience of his own).

"The whole ordeal was incredibly challenging, and yet somehow we managed to make it work," says Jeff. "On the very day we moved in, with the moving trucks outside the house, I had to run a wheel of cheese over to a chef with a standing order."

Another hitch: The entire cheese-making area was a 350-square-foot converted mudroom off the back of the farmhouse. Jeff says it reached a point where the cost of upgrading it outweighed the convenience of staying put. When their current property came on the market, in a neighboring province, they couldn't resist — the local government had been trying to grow opportunities for makers of cheese, charcuterie, and craft beer to drive up tourism in this popular vacation spot.

That meant Jeff was able to offset some of the building expenses through grants (not the easiest of processes) with the stipulation that he renovate an existing structure. "It was a dilapidated barn with an old tractor, oil leaks on the floor, and tons of pigeons," he recalls. "It would have been easier to just design from scratch."

As for the cheeses, Jeff says he wanted to cover the spectrum — including fresh (Flower Meadow), blue (Highland Blue), bloomy white (Madawaska), and their newest offering, a washed-rind cheese called Ompah (all names are drawn from the surrounding area), which took him about two years to get right and has already become popular with his customers. For now, his most pressing concern is spending time with his newborn daughter. "Sure hope she likes cheese."

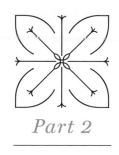

Part 2

RECIPES

FOR ALL TYPES OF CHEESE

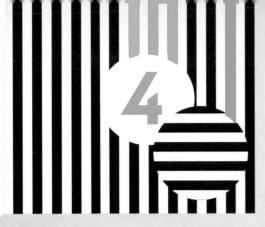

FRESH CHEESES

Fresh cheeses require little equipment and are excellent choices for beginning cheese makers, as they are quick, delicious, and easy to make — no fermenting or aging required. They are also perfect for experimentation because you can vary them simply by adding herbs, spices, honey, or other flavorings (see the box on page 73 for suggestions).

Most of these high-moisture cheeses have a creamy, spreadable consistency. Many are called bag cheeses because the curds are drained in a "bag" of butter muslin. They are made by coagulating milk or cream with cheese starter or with an acid, such as vinegar or lemon juice. Some recipes call for a little rennet to help firm the curds.

It's important to drain these cheeses in a place where the temperature stays close to 72°F (22°C) — usually the kitchen. If the temperature and humidity are too high, you will have problems with yeast, which may produce a gassy, off-flavored cheese. If the temperature is too low, the cheese will not drain properly. The yield from 1 gallon of milk is usually 1½ to 2 pounds of soft cheese, depending on the type of milk you use and the desired consistency of the cheese. The greater the butterfat content, the higher the cheese yield.

30-MINUTE MOZZARELLA,
page 92

Fresh cheeses will keep for one to two weeks in the refrigerator. Although it is not my first choice, they may also be frozen. If you want to salt your cheese, it's best to wait until after thawing to do this, as salt will increase the freezing temperature of the cheese and therefore it will not keep as well.

The techniques used in this section are very straightforward; see chapter 3 to review them as needed.

 Recipes with this icon call for warming the milk directly on the stove over medium heat.

 Recipes with this icon call for heating the milk indirectly, with the cheese pot resting in a bowl or sink full of hot water (see page 37 for more instruction).

Recipes that can be made using either method have both icons.

RICOTTA, *page 88*

LACTIC CHEESE

*This delicious spreadable cheese is easy to make and ready to eat in just
24 hours. I like to make this cheese at night and drain it in the morning. You
may add herbs and spices in a variety of combinations for truly tantalizing
variations. I find that a combination of freshly ground black pepper, minced
garlic, chopped fresh chives, and a dash of paprika or pimentón makes a tasty
savory cheese. You can also add honey or another sweetener as desired for
using it in desserts.*

1 gallon milk

¼ teaspoon calcium
chloride diluted
in ¼ cup cool,
nonchlorinated water

1 packet direct-set
mesophilic starter
culture

3 drops liquid rennet
diluted in ⅓ cup cool,
nonchlorinated water

Cheese salt (optional)

Yield: About 2 pounds

1. Heat the milk to 86°F (30°C). Add the calcium chloride solution and stir
   well to combine. Sprinkle the starter over the surface of the milk, wait
   2 minutes for the powder to rehydrate, then stir well.

2. Add 1 teaspoon of the diluted rennet and stir gently with an up-and-
   down motion for 30 seconds. Cover and let set undisturbed at 72–86°F
   (22–30°C) for 10–12 hours, or until a solid curd forms. The curd will look
   like a thick yogurt.

3. Gently ladle the curd into a colander lined with butter muslin. Tie
   the corners of the muslin into a knot and hang the bag to drain for
   6–12 hours, or until the cheese reaches your desired consistency. A room
   temperature of at least 72°F (22°C) will encourage proper drainage. If you
   want the curds to drain more quickly, change the muslin periodically.

4. Remove the cheese from the bag and place in a bowl. Add salt to taste, if
   using. Cover the bowl and store it in the refrigerator for up to 2 weeks.

TROUBLESHOOTING

If the cheese has a hard, rubbery texture, add less rennet next time. If the
cheese is too moist, add a little more rennet or drain longer.

Fromage Blanc

French in origin, fromage blanc simply means "white cheese." The word fromage is derived from the Greek word formos, *the name for the wicker baskets used by the Greeks to drain whey from cheese. Fromage blanc is easy to prepare and makes an excellent spread, with the consistency of cream cheese (and only a fraction of the fat and calories). Add herbs and spices to it, or use it plain as a substitute for cream cheese or ricotta in cooking. Fromage blanc can be made with either whole or skim milk; if you use the latter, the yield will be lower and the cheese will be drier. I like to make this cheese at night so I can drain it when I get up in the morning.*

1 gallon milk

¼ teaspoon calcium chloride diluted in ¼ cup cool, nonchlorinated water

1 packet direct-set fromage blanc starter culture

Yield: About 2 pounds

1. Heat the milk to 86°F (30°C). Add the calcium chloride solution and stir well to combine. Sprinkle the starter over the surface of the milk, wait 2 minutes for the powder to rehydrate, then stir well.

2. Cover and let set at 72–86°F (22–30°C) for 10–12 hours. The curd mass should pull slightly away from the edge of the pot, with a small amount of clear whey visible.

3. Gently ladle the curds into a colander lined with butter muslin. Tie the corners of the muslin into a knot and hang the bag to drain at 72°F (22°C) for 4–6 hours, or until the cheese reaches your desired consistency. A shorter draining time produces a thinner, more spreadable cheese. A longer draining time produces a cream cheese–type consistency.

4. Place the cheese in a covered container and store in the refrigerator for up to 2 weeks.

   NOTE: Fromage Blanc made with skim milk duplicates a delicious variety of the German-style Quark on page 91.

MASCARPONE WITH CULTURE

Very similar to cream cheese, mascarpone is an Italian soft coagulated cream used in cooking and desserts such as Italian pastries and tiramisu. This recipe produces a rich, velvety texture and sweet flavor. Mix mascarpone with herbs and spices or serve it plain with dried figs and apricots. One of my favorite ways to serve it is to mix it with crumbled blue cheese for a delicious dip. If you are going to a party on Saturday and start this cheese Thursday night, you will have a fantastic cheese to bring with you.

1 quart light cream or half-and-half

¼ teaspoon calcium chloride diluted in ¼ cup cool, nonchlorinated water

1 packet direct-set crème fraîche starter culture

Yield: About 1 pound

1. Heat the cream to 86°F (30°C). Add the calcium chloride solution and stir well to combine. Sprinkle the starter over the surface of the milk, wait 2 minutes for the powder to rehydrate, then stir well.

2. Cover and let set undisturbed at 72°F (22°C) for 10–12 hours, or until fully coagulated.

3. Ladle the curds into a colander line with butter muslin and drain at 72°F (22°C) for 3–4 hours. If a thicker curd is desired, continue draining in the refrigerator for 1–2 hours longer, depending on your desired consistency.

4. Place the cheese in a covered container and store in the refrigerator for up to 2 weeks.

   NOTE: If this cheese comes out too firm for you, put it in a food processer and give it a quick whirl.

Mascarpone
with Tartaric Acid

Famous as the ultradelicious filling for cannolis, this version of mascarpone also makes a splendid spread for slices of sweet bread. The consistency is similar to that of butter, depending on the amount of time it drains. Tartaric acid is a natural vegetable acid derived from the seed of the tamarind tree, which is found throughout the Caribbean.

2 ice cubes

1 quart light cream or half-and-half

¼ teaspoon calcium chloride diluted in ¼ cup cool, nonchlorinated water

⅛–¼ teaspoon tartaric acid diluted in ½ cup cool, nonchlorinated water

Yield: About 1 pound

1. Roll the ice cubes around the pot until it feels cold to the touch from the outside (see page 70). Add the milk.

2. Add the calcium chloride solution to your milk and stir well to combine, without touching the bottom of the pot with your ladle. Heat the cream to 185°F (85°C).

3. Add the diluted tartaric acid and stir slowly for 30 seconds. The mixture will slowly thicken into a Cream of Wheat consistency, with tiny flecks of curd. If the cream does not coagulate, increase the heat to 200°F (93°C). If it still won't set, add a speck more of tartaric acid solution and stir. Be careful not to add too much, or a grainy texture and an acidic flavor may result.

4. Ladle the curds into a colander lined with butter muslin and drain at 72–86°F (22–30°C) for 1 hour.

5. Place the cheese in a covered container and store in the refrigerator for up to 2 weeks.

TROUBLESHOOTING
If the cheese does not set properly, it may be the cream. Ultrapasteurized cream often does not work.

FROMAGE BLANC,
page 63

MASCARPONE WITH
TARTARIC ACID, *page 65*

MOIST BUTTERMILK CHEESE

This cheese is fairly moist, because the curds are not heated before draining. Use fresh, homemade, cultured buttermilk made from a packet of buttermilk starter. It will have a thick, clabbered consistency (see page 263 for the recipe).

1 quart freshly made buttermilk

1 drop liquid rennet diluted in 2 tablespoons of cool, nonchlorinated water, (only if using goat's-milk buttermilk)

Cheese salt (optional)

Herbs (optional)

Yield: 6–8 ounces

1. Allow the buttermilk to set for 24 hours at 72°F (22°C). This will lightly sour it. (If you prefer a less sour cheese, omit this step.) If you are using buttermilk made with goat's milk, add 1 teaspoon of the diluted rennet to help it set.

2. Gently ladle the buttermilk into a colander lined with butter muslin. Tie the corners of the muslin into a knot and hang the bag to drain at 72–86°F (22–30°C) for 12 hours, or until the cheese reaches your desired consistency.

3. Remove the cheese from the bag and place in a bowl. Add salt and herbs to taste, if using. Cover the bowl and store it in the refrigerator for up to 2 weeks.

Dry Buttermilk Cheese

Because this buttermilk cheese is heated, its texture is fairly dry, with a grainy, spreadable consistency and a slightly acidic flavor. For best results, use fresh, homemade, cultured buttermilk (see page 263).

2 ice cubes

1 quart buttermilk

Cheese salt (optional)

Herbs (optional)

Yield: 6–8 ounces

1. Roll the ice cubes around the pot until it feels cold to the touch from the outside (see page 70). Add the buttermilk.

2. Heat the buttermilk to 160°F (71°C), stirring occasionally. The curds will separate from the whey; if they don't, increase the temperature to 180°F (82°C).

3. Gently ladle the curds into a colander lined with butter muslin. Tie the corners of the muslin into a knot and hang the bag to drain at 72–86°F (22–30°C) for 6–10 hours, or until the curds reach your desired consistency.

4. Remove the cheese from the bag and place in a bowl. Add salt and herbs to taste, if using. Cover the bowl and store in the refrigerator for up to 2 weeks.

THE ICE-CUBE TRICK

I use the following tip whenever I make an acid-set or high-temperature cheese. Roll a couple of ice cubes around the bottom of your pot until slightly melted, then add your milk. The ice creates a liquid barrier that prevents scorching and helps with cleanup. It's not necessary to discard the ice cubes; the amount of water they add is miniscule compared to the water in milk.

With this method, I find that I don't have to stir the milk until it reaches 180°F (82°C) and only minimal stirring is required after that. The less stirring, the better the curd with these fresh cheeses! When you do stir, be sure that you don't touch the bottom of the pot with the ladle, which will break the barrier.

LEMON CHEESE

This moist cheese has a spreadable consistency and a mild, lemony flavor.

2 ice cubes

½ gallon milk

¼ teaspoon calcium
chloride diluted
in ¼ cup cool,
nonchlorinated water

Juice of 2–3 large
lemons (about ¾ cup)

Cheese salt (optional)

Herbs (optional)

Yield: About 1 pound

1. Roll the ice cubes around the pot until it feels cold to the touch from the outside (see page 70). Add the milk.

2. Add the calcium chloride solution and stir well to combine.

3. Heat the milk to 175°F (79°C). Add two-thirds of the lemon juice and stir well to combine without touching the bottom of the pot with your ladle.

4. Cover and let set for 15 minutes. You are looking for a clear separation of the curds and whey, not milky whey. If the milk does not separate, increase the heat to 190°F (88°C). If it still won't separate, stir in more lemon juice a little at a time until it separates.

5. Gently ladle the curds into a colander lined with butter muslin. Tie the corners of the muslin into a knot and hang the bag to drain at 72–86°F (22–30°C) for 1–2 hours, or until the cheese reaches your desired consistency.

6. Remove the cheese from the bag and place in a bowl. Add the salt and herbs to taste, if desired. Cover the bowl and store in the refrigerator for up to 2 weeks.

   NOTE: For a refreshing summertime drink, drain the curds for just 20 minutes, then chill the mixture and add mint leaves.

Neufchâtel

Originally from the town of Neufchâtel in Normandy, France, this cheese is made from whole milk enriched with cream. Americans tend to eat it fresh; the French prefer it ripened with a white surface mold. Once you perfect this recipe, try seasoning it with chives, garlic, onions, scallions, olives, or cornichons.

1 gallon milk

1 pint heavy cream

¼ teaspoon calcium chloride diluted in ¼ cup cool, nonchlorinated water

1 packet direct-set mesophilic starter culture

3 drops liquid rennet diluted in ⅓ cup cool, nonchlorinated water

Cheese salt (optional)

Herbs (optional)

Yield: About 2 pounds

1. Combine the milk and cream. Heat to 80°F (27°C). Add the calcium chloride solution and stir well to combine. Sprinkle the starter over the surface of the milk, wait 2 minutes for the powder to rehydrate, then stir well.

2. Add 1 teaspoon of the diluted rennet. You may have to adjust the precise amount of rennet after experimenting with your milk. The exact amount of rennet is important. Too little, and the cheese will drain through the muslin; too much, and the cheese will be hard and rubbery. Stir slowly with an up-and-down motion for 30 seconds.

3. Cover and let set at 72–86°F (22–30°C) for 12–18 hours, or until a thick curd has formed. It will look like thick yogurt.

4. Gently ladle the curds into a colander lined with butter muslin. Tie the corners into a knot and hang the bag to drain at 72–86°F (22–30°C) for 6–12 hours, or until the cheese reaches your desired consistency.

5. Put the bag into a colander lined with butter muslin and place the colander in a pot. Put a plate on the bag and place a weight on the plate (the equivalent of two bricks is sufficient). Cover the pot and refrigerate for 13 hours.

6. Remove the cheese from the bag and put it in a bowl. Add salt and herbs to taste, if using, and mix well. Roll the cheese briefly by hand until it holds together, then divide it into four parts. Shape each part into a round.

7. Wrap each round separately in cheese wrap. Store in the refrigerator for up to 2 weeks.

Flavoring Fresh Cheeses

The possibilities for incorporating sweet or savory flavors into your fresh cheeses are endless. Use these as starting points and then experiment with your own favorite herbs, spices, and other seasonings.

Always add these after the cheese has drained. Add small amounts at first so your cheese is not overwhelmed by the flavoring. Keep in mind, you can always add more but you can't take any out. Rather than mixing a flavoring in, you can also form your finished cheese into logs or balls and coat them with the flavoring.

- Chopped dried or fresh herbs, such as mint, parsley, basil, chives, dill, thyme, and rosemary

- Minced garlic, shallot, or spring onion

- Grated fresh horseradish or wasabi paste
- Freshly ground pepper or crushed whole peppercorns
- Red pepper flakes or crumbled or ground dried chile peppers
- Toasted cumin, mustard, fennel, or sesame seeds
- Za'atar, ras el hanout, or other spice blends
- Ground spices such as cardamom, cloves, cinnamon, allspice, nutmeg, and coriander
- Poppy seeds
- Dried lemon or orange peel
- Chocolate chips
- Coconut
- Honey

GERVAIS

Gervais is a French cheese made from a mixture of milk and cream. It is similar to Neufchâtel but richer and creamier, and rarely found in cheese stores. It is traditionally made with cow's milk but you may use goat's milk.

2⅓ cups milk

1⅓ cups heavy cream

¼ teaspoon calcium chloride diluted in ¼ cup cool, nonchlorinated water

1 packet direct-set mesophilic starter culture

1 drop liquid rennet diluted in 2 table-spoons cool, non-chlorinated water

Cheese salt

Herbs (optional)

Yield: 8–10 ounces

1. Combine the milk and cream. Warm to 80°F (27°C). Add the calcium chloride solution and stir well to combine. Sprinkle the starter over the surface of the milk, wait 2 minutes for the powder to rehydrate, then stir well.

2. Add the diluted rennet and stir gently with an up-and-down motion for 30 seconds. If the milk starts to coagulate, stop stirring. Cover and let set at 72–86°F (22–30°C) for 24 hours.

3. Gently ladle the curds into a colander lined with butter muslin. Tie the corners of the muslin into a knot and hang the bag to drain at 72–86°F (22–30°C) for 4–6 hours, or until the cheese reaches your desired consistency. You may need to take down the bag periodically and scrape the sides with a spoon to open the pores of the muslin for better drainage.

4. Remove the cheese from the muslin and place it in a 1- to 2-pound cheese mold lined with butter muslin. Press at 15 pounds of pressure for 6–8 hours.

5. Remove the cheese from the mold and transfer it to a bowl. Add a pinch of salt to taste and herbs if desired; mix well. The cheese will be creamy and smooth. If it's too lumpy, blend in a food processor for 1 minute.

6. To store this cheese, pack it into small molds lined with wax paper. Traditionally, these molds are 2¼ inches wide by 1¾ inches deep. This recipe will make four cheeses. Place the filled molds in a covered container and store in the refrigerator; the cheese tastes best when eaten within a few days but will keep for up to 2 weeks.

BONDON

Bondon is another fresh French cheese made from whole milk and is similar in taste and texture to Neufchâtel. Traditionally, this cheese is made from cow's milk, but you can use goat's milk with this recipe.

1 quart milk

¼ teaspoon calcium chloride diluted in ¼ cup cool, nonchlorinated water

1 packet direct-set mesophilic starter culture

1 drop liquid rennet diluted in 2 tablespoons cool, nonchlorinated water

Cheese salt (optional)

Herbs (optional)

Yield: About 6 ounces

1. Warm the milk to 65°F (18°C). Add the calcium chloride solution and stir well to combine. Sprinkle the starter over the surface of the milk, wait 2 minutes for the powder to rehydrate, then stir well.

2. Add the diluted rennet and stir gently with an up-and-down motion for 30 seconds. Cover and let set at 65°F (18°C) for 24 hours, or until coagulated.

3. Gently ladle the curds into a colander lined with butter muslin. Tie the corners of the muslin into a knot and hang the bag to drain at 72–86°F (22–30°C) for 6–8 hours, or until the cheese reaches your desired consistency. You may need to scrape the muslin with a spoon occasionally to hasten draining.

4. Remove the cheese from the muslin and place it in a 1- to 2-pound cheese mold lined with butter muslin. Press the cheese at 15 pounds of pressure for 4–8 hours.

5. Remove the cheese from the mold and put it in a bowl. Add the salt and herbs to taste, if using. The cheese will be smooth in texture. If the texture is too grainy or firmer than you would like it, blend in a food processor for 1 minute.

6. To store this cheese, pack it into one or two small molds lined with wax paper. Traditionally, these molds are 1¾ inches wide by 2¾ inches deep. Place the filled molds in a covered container and store in the refrigerator for up to 2 weeks.

NOTE: If you use 1 gallon of milk, you do not have to change the amount of starter for this cheese.

Cream Cheese:
Uncooked–Curd Method

Rich, creamy, and easy enough for kids to make and enjoy.

2 quarts light cream or half-and-half

¼ teaspoon calcium chloride diluted in ¼ cup cool, nonchlorinated water

1 packet direct-set mesophilic starter culture

Cheese salt (optional)

Herbs (optional)

Yield: About 1 pound

1. Heat the cream to 86°F (30°C). Add the calcium chloride solution and stir well to combine. Sprinkle the starter over the surface of the milk, wait 2 minutes for the powder to rehydrate, then stir well.

2. Cover and let set at 86°F for 8–12 hours, or until a solid curd forms.

3. Gently ladle the curds into a colander lined with butter muslin. Tie the corners of the muslin into a knot and hang the bag to drain at 72–86°F (22–30°C) for 8–12 hours, or until the cheese reaches your desired consistency. Scraping the sides of the bag once or twice will speed the draining process.

4. Place the cheese in a bowl and add the salt and herbs to taste, if using.

5. Pack the cheese into small molds and place in the refrigerator until firm, a few hours. Take them out of the molds and wrap individually in cheese wrap. Store in the refrigerator for up to 2 weeks.

CREAM CHEESE:
COOKED-CURD METHOD

This recipe requires a bit more work and produces a slightly drier cheese than the uncooked method.

2 quarts light cream or half-and-half

¼ teaspoon calcium chloride diluted in ¼ cup cool, nonchlorinated water

1 packet direct-set mesophilic starter culture

3 drops liquid rennet diluted in ⅓ cup cool, nonchlorinated water

1–2 quarts water

Cheese salt (optional)

Herbs (optional)

Yield: About 1 pound

1. Heat the cream to 86°F (30°C). Add the calcium chloride solution and stir well to combine. Sprinkle the starter over the surface of the milk, wait 2 minutes for the powder to rehydrate, then stir well.

2. Add 1 teaspoon of the diluted rennet and stir gently with an up-and-down motion for 30 seconds. Cover and let set at 72–86°F (22–30°C) for 10–12 hours, or until the curd is fully set.

3. Heat the water to 170°F (77°C). Add enough of the hot water to the curd to raise its temperature to 125°F (52°C).

4. Gently ladle the curd into a colander lined with butter muslin. Tie the corners of the muslin into a knot and hang the bag to drain at room temperature for 8–12 hours, or until the cheese reaches your desired consistency.

5. Place the cheese in a bowl and add salt and herbs to taste, if using.

6. If you would like to shape your cheese, place it into two or three individual small molds and chill in the refrigerator for 3 hours. Once the cheeses are firm, unmold them and wrap individually in cheese wrap. Store in the refrigerator for up to 2 weeks.

CREOLE CREAM CHEESE

Created by the French settlers in Louisiana back in the 1800s, Creole cream cheese remains a New Orleans tradition. The name is misleading, as the cheese is more akin to farmer's cheese that's been crossed with sour cream, with buttermilk lending subtle tang. The cheese was originally sold in pint containers as one big curd, topped with heavy cream, and eaten for breakfast with sugar; it's also heavenly on toast and topped with fruit or jam. The following is reportedly the best recorded recipe. It was written by Myriam Guidroz and appeared in the *Times-Picayune* newspaper.

1 gallon skim milk (may be reconstituted dry milk powder)

½ cup cultured buttermilk

½ teaspoon liquid rennet

Half-and-half or heavy cream

YIELD: 1–2 POUNDS

1. Make sure the temperature of the milk is no cooler than 70°F (21°C) and no warmer than 80°F (27°C). Place the milk in a large container.

2. Add the buttermilk and stir well. Add the rennet and agitate vigorously for 1 minute. Cover and let set at room temperature (72°F/22°C) for 12–15 hours.

3. After the cheese has set, ladle it into Creole cream cheese molds (or other containers with perforated bottoms, such as heart-shaped molds or plastic butter tubs with holes punched in them so that the whey can drain). In a large roasting pan, elevate a rack with custard cups, then place the molds on the rack. Refrigerate until no more whey drips out. The cheese will take at least 4–6 hours to form.

4. Unmold the cheese and store in covered containers in the refrigerator for at least 1 month. To eat it, spoon the amount you want into a bowl and cover with half-and-half.

SWISS–STYLE CREAM CHEESE

This recipe makes a sweet cream cheese that is utterly delectable.

1 quart heavy cream

¼ teaspoon calcium chloride diluted in ¼ cup cool, nonchlorinated water

1 packet direct-set mesophilic starter culture

1 drop liquid rennet diluted in 2 tablespoons cool, nonchlorinated water

2 teaspoons cheese salt

Herbs (optional)

Yield: About 8 ounces

1. Warm the cream to 65°F (18°C). Add the calcium chloride solution and stir well to combine. Sprinkle the starter over the surface of the milk, wait 2 minutes for the powder to rehydrate, then stir well.

2. Add the diluted rennet and stir gently with an up-and-down motion for 30 seconds. Cover and let set at 65°F (18°C) for 14–16 hours.

3. Gently ladle half of the curds into a colander lined with butter muslin. Sprinkle with 1 teaspoon of the salt. Ladle the remaining curds into the colander and sprinkle with the remaining 1 teaspoon salt. (The salt will help the cream drain better.) Tie the corners of the muslin into a knot and hang the bag to drain at 72–86°F (22–30°C) for 12 hours.

4. Place the curd into a mold lined with cheesecloth and press for 4–6 hours at 10 pounds of pressure.

5. Remove the cheese from the mold, packing it into small containers, cover, and store in the refrigerator for up to 1 week. If adding herbs, place the cheese in a bowl and mix them in before dividing it into the containers.

Liam Callahan
BELLWETHER FARM
Petaluma, California

Liam and Cindy Callahan

BELLWETHER FARM IS YET ANOTHER STORY of cheese making as a result of "why nots" and "what ifs." In 1986, city transplants Cindy and Ed Callahan purchased 34 acres in Petaluma, about 45 miles from San Francisco. Then Cindy bought 25 sheep to help mow the lawn — and ended up with too much lamb for one family to eat when the animals reached retirement. "So I contacted area restaurants and Chez Panisse was one of the first to respond. That's how Bellwether Farm got started," explains Cindy's son Liam, who was then attending nearby UC Berkeley.

The transition to cheese came a couple years later, when a Syrian friend suggested making yogurt or cheese from sheep's milk like they did back home. "We were stupefied to realize our favorite cheeses, especially Roquefort, were made with sheep's milk. Almost no one was making sheep's-milk cheese in the United States back then." Fresh out of college, Liam joined his mom in launching the creamery. "We found some books, mastered some basic methods, and made a fresh, bag-drained, long-set cheese, which we put in jelly jars and took to the farmers' markets."

Because sheep give very little milk, the Callahans had to freeze it to collect enough to make cheese. "We were fortunate to borrow a cheese room at a nearby creamery on the weekends and make small batches in their 20-gallon vat." Once their fresh cheese was a success, they moved on to aged cheeses and also started buying in cow's milk from neighboring farms to make traditional Italian cheeses, including basket-dipped ricotta. But the 300 sheep — and their cheeses — are still the stars.

Liam remains the main cheese maker, working the vats, adding the rennet, and filling out the food-safety forms, plus overseeing all the other moving pieces of the dairy and creamery. At some point, however, he wanted to be able to do more than just make and sell cheese. "There are so many problems with food availability, with people understanding what constitutes a healthy meal. So we created the Bellwether Farms Foundation, pledging 1 percent of all sales to food-related organizations involved in hunger relief or education. All of our new packaging has our foundation logo on it, which is an illustration based on a photo of my mom holding a lamb — a nod to how Bellwether all began."

FRENCH–STYLE CREAM CHEESE

This version of cream cheese is surprisingly sweet and deliciously creamy.

2 cups heavy cream

2 cups milk

¼ teaspoon calcium chloride diluted in ¼ cup cool, nonchlorinated water

1 packet mesophilic direct-set starter culture

1 drop liquid rennet diluted in 2 tablespoons cool, nonchlorinated water

Cheese salt (optional)

Herbs (optional)

1–2 tablespoons heavy cream (optional)

Yield: About 1 pound

1. Combine the cream and milk. Heat the mixture to 86°F (30°C). Add the calcium chloride solution and stir well to combine.

2. Sprinkle the starter over the surface of the milk, wait 2 minutes for the powder to rehydrate, then stir well. Add the diluted rennet and stir gently with an up-and-down motion for 30 seconds. Cover and allow to set at 72–86°F (22–30°C) for 24 hours.

3. Gently ladle the curds into a colander lined with butter muslin. Tie the corners of the muslin into a knot and hang the bag to drain at 72–86°F (22–30°C) for 6–12 hours, or until the cheese reaches your desired consistency.

4. Place the curds in a bowl and mix by hand to a pastelike consistency. Add the salt and herbs to taste, if using. For a creamier cheese, add the optional heavy cream.

5. Pack the cheese into a container, cover, and store in the refrigerator for up to 1 week.

TROUBLESHOOTING
If you find that the heavy cream will not set, use light cream or half-and-half next time.

Whole-Milk Ricotta

Traditionally, ricotta is made by reheating the whey after making cheese from ewe's milk. To make ricotta from whey, see Ricotta with Whey and Milk (page 257) and Pure Whey Ricotta (page 255). This simple variation uses whole milk from the grocery store instead of whey; the resulting ricotta has a good flavor and a high yield. To take it a step further and make a grating cheese, see Ricotta Salata (page 179).

2 ice cubes

1 gallon whole milk

1 teaspoon citric acid dissolved in 1 cup cool, nonchlorinated water

1 teaspoon cheese salt dissolved in ½ cup cool, nonchlorinated water

1–2 tablespoons heavy cream (optional)

Yield: 1½–2 pounds

1. Roll the ice cubes around the pot until it feels cold to the touch from the outside (see page 70).

2. Pour in the milk, add the citric acid and the salt solutions, and mix well without touching the bottom of the pot.

3. Heat the milk to 185–195°F (85–91°C); do not boil. At 180°F (82°C), stir occasionally to prevent a skim coat from forming on top of the milk.

4. As soon as the curds and whey separate (the whey should be translucent, not milky), turn off the heat. If the whey is still milky at 195°F, increase the heat another 10–20°F (5–10°C). If the whey remains milky, add more citric acid solution a little at a time, stirring gently until you see a clear separation. Allow to set undisturbed for 10 minutes.

5. Gently ladle the curds with a slotted spoon into a colander lined with butter muslin. Tie the corners of the muslin into a knot and hang the bag to drain at 72–86°F (22–30°C) for 20–30 minutes, or until the cheese reaches your desired consistency.

6. Remove the cheese from the bag and place in a covered container. For a creamier consistency, add the cream at the end and mix thoroughly. Store in the refrigerator for up to 2 weeks.

LARGE-CURD COTTAGE
CHEESE, *page 113*

RICOTTA

CREAM CHEESE (UNCOOKED-
CURD METHOD), *page 76*

SCHIZ

This wonderfully simple and delicious cheese hails from the Dolomites of Italy. The cheese, like the place that it was created in, is something of an undiscovered hidden gem. Try frying the finished cheese to serve over polenta.

1 gallon milk

¼ teaspoon calcium chloride diluted in ¼ cup cool, nonchlorinated water

¼ teaspoon liquid rennet (or ¼ rennet tablet) diluted in ¼ cup cool, non-chlorinated water

Yield: 1 pound

1. Slowly heat the milk to 96°F (36°C). Add the calcium chloride solution and stir well to combine.

2. Add the diluted rennet and stir gently with an up-and-down motion for 30 seconds. Cover and let set undisturbed at 96°F for 30 minutes, or until the curd gives a clean break.

3. Cut the curds vertically into ¾-inch columns. Let the whey rise from the cuts for 2–3 minutes. Then use a ladle or a spoon to cut through the columns, creating small cubes of curd.

4. Over the next 20–30 minutes, stir intermittently, keeping the curds at 96°F (36°C). The longer you cook the curds, the drier the cheese will be. The final result will be firm curds, fully cooked through and resistant to the touch. Remove the pot from the heat and let the curds settle in the whey. Remove enough whey from the pot so that the liquid is just above the curd level.

5. Use a ladle to transfer the curd to a basket mold. Let the cheese drain at room temperature (72–86°F/22–30°C), flipping it every 30 minutes, until the drainage slows, 3–4 hours. Store in a covered container in the refrigerator and eat within 2 days.

   NOTE: To make this cheese with raw milk, reduce the rennet to ⅛ teaspoon.

INDIA MAY BE THE LARGEST PRODUCER OF MILK in the world, but cheese (beyond paneer) has never been part of the country's culture. "When I made an authentic pasta filata mozzarella in 1994," says Mukund Naidu, "very few locals had a clue as to what I was talking about."

In the years since then, Mukund has seen slow and steady progress. "People travel to other countries and come back wanting to sample the cuisines they savored on their trips. Plus, the opening up of trade barriers has allowed gourmet cheeses to be more readily available in the major metro areas."

Mukund's own cheese-making journey began when he fled the city as a mechanical engineering student in 1991 to join a farm in Kodaikanal, in the mountains of southern India, that primarily grew coffee beans. Milk was another source of income; as the size of the dairy herd grew, the owner (an Italian-American expat) started making cheese to sustain the farm, bringing samples from around the world to learn more. Mukund's aha moment was the day he tasted Roquefort cheese, after overcoming his fear of eating all that unfamiliar mold. "I was astounded by the flavors. I thought, 'Wow, if this is cheese, then this is what I need to make for the rest of my life.'"

In 2001, Makund left to start his own small operation in Goa, on the west coast, specializing in that authentic mozzarella along with feta, ricotta, and mascarpone, plus a small amount of fresh cheese for hotels in Mumbai. Some 15 years later, he still operates out of the same facility and faces the same challenges of cheese makers near and far — notably finding quality milk, as well as the costs and logistics of proper shipping and competition from imported cheese products.

Besides being among the first successful artisanal cheese makers (he estimates there are now around 30 across India), Mukund devotes much of his time to training others, in India and elsewhere. "If they're looking for big bucks, I tell them this is the wrong profession. But if they're looking at this as a passion, there is nothing more satisfying. Cheese has been called 'milk's leap toward immortality.' Milk spoils; its shelf life is short. But with cheese, you create something living and exciting; there's chemistry inside."

A CHEESE MAKER'S STORY

Mukund Naidu
THE CREAMERY
Bangalore, India

Paneer

Paneer (or panir) is an Indian cheese similar to farmer's cheese — and is one of the simplest types of unripened cheeses to make. It is rather mild and readily absorbs the flavors of the seasonings it is cooked with.

2 ice cubes

1 gallon whole milk

1 teaspoon calcium chloride diluted in 1 cup cool, nonchlorinated water

½ cup lemon juice or 1 teaspoon citric acid dissolved in 2 cups hot water (170°F/77°C)

Cheese salt (optional)

Yield: 1¾–2 pounds

1. Roll the ice cubes around the bottom of the pot until it feels cold to the touch from the outside (see page 70).

2. Add the milk and the calcium chloride solution and stir well to combine. Heat to 195°F (91°C) and hold for 20–30 minutes.

3. Cool the milk to 170°F (77°C). Add either the lemon juice or the diluted citric acid. Stir slowly until the curd is separated.

4. Once the curds clearly separate from the whey, remove from the heat and let rest for 10–20 minutes.

5. Ladle the curds into a colander lined with butter muslin. Allow the curds to drain for 30 minutes.

6. Tie the corners of the cloth together to make a bag and leave it in the colander. Place a plate on top of the bag and add a 15-pound weight. Press for 15 minutes (longer if a drier cheese is desired).

7. Remove the cheese from the bag. If using salt, rub just under a tablespoon of salt on the surface and let it absorb into the cheese. Traditionally, paneer is eaten within a day, but it can be stored in a covered container in the refrigerator for up to 7 days unsalted and up to 14 days if lightly salted.

CHENNA

Essentially the same cheese as paneer, chenna is kneaded while still warm into a light, velvety smooth, whipped-cream consistency. It is an essential ingredient in many Bengali sweets.

1. Follow the recipe for paneer through step 5.

2. Return the wrapped cheese to the colander, place a 5-pound weight on top, and press for 45 minutes.

3. Remove the still-warm cheese from the bag and place on a smooth, clean work surface. Break it apart and press with a clean cloth to remove any remaining whey.

4. Knead the cheese by pressing out with the heel of your palm and the flat of your hand. Gather up the cheese with a spatula and continue kneading for up to 10 minutes, or until the cheese is light and velvety smooth, without any grainy texture.

5. Place the cheese in a covered container and store in the refrigerator for up to 2 weeks.

BENGALI RASGULLA

Rasgulla, a popular dessert made from chenna, are lighter-than-air balls that melt in your mouth. Simply form the chenna into uniform balls, about ½ inch in diameter (they will double in size during cooking), while bringing 2 parts water to 1 part sugar to a boil in a wide pan or deep skillet to form the poaching syrup.

Once the syrup is clear and the sugar has dissolved, add a bit of rose water and/or cracked cardamom pods, if desired, and then the chenna balls, cooking in batches so they have plenty of room to expand. Cover and cook at a gentle boil until the dumplings are cooked through, about 10 minutes, stirring occasionally.

To test for readiness, drop a cheese ball in a glass of tap water; it should sink to the bottom. If it doesn't, continue cooking. Remove with a slotted spoon. Serve chilled or at room temperature.

Queso Blanco

Queso blanco ("white cheese") is a firm cheese, with a bland, mildly sweet flavor. It is extremely easy to make and an excellent choice if you are in a hurry or if the weather is very hot, a condition that causes problems in the production of many cheeses.

Because it has the unusual property of not melting even when deep-fried, queso blanco is one fresh cheese that is excellent for cooking — usually diced into ½-inch cubes. It also browns nicely and takes on the flavor of the food and spices it is cooked with.

2 ice cubes

1 gallon milk

¼ cup vinegar
(5% acidity)

Yield: 1½–2 pounds

1. Roll the ice cubes around the pot until it feels cold to the touch from the outside (see page 70). Add the milk.

2. Heat the milk to 185–190°F (85–88°C).

3. Slowly add the vinegar, a little at a time, until the curds separate from the whey. Usually ¼ cup of vinegar will precipitate 1 gallon of milk. If you still have milky whey, increase the temperature to 200°F (93°C). (Do not boil; this will result in a "cooked" flavor.) If the whey is still milky, add a bit more vinegar.

4. Gently ladle the curds into a colander lined with butter muslin. Tie the corners of the muslin into a knot and hang the bag to drain at 72–86°F (22–30°C) for 1–2 hours, or until the cheese reaches your desired consistency.

5. Remove the cheese from the bag and place it in a covered container. Store in the refrigerator for up to 1 week.

QUESO FRESCO

This Latin American quick farm cheese may be made using a variety of methods. The following technique is simple to do at home.

2 gallons milk

¼ teaspoon calcium chloride diluted in ¼ cup cool, nonchlorinated water

1 packet direct-set mesophilic starter culture

¼ teaspoon liquid rennet (or ¼ rennet tablet) diluted in ¼ cup cool, nonchlorinated water

2 tablespoons cheese salt

Yield: 2 pounds

1. Heat the milk to 90°F (32°C). Add the calcium chloride solution and stir well to combine. Sprinkle the starter over the surface of the milk, wait 2 minutes for the powder to rehydrate, then stir well.

2. Add the diluted rennet and stir gently with an up-and-down motion for 30 seconds.

3. Cover and let set for 30–45 minutes, or until the curd gives a clean break.

4. Cut the curd into ¼-inch cubes.

5. Over the next 20 minutes, gradually increase the temperature to 95°F (35°C), stirring gently every few minutes to keep the curds from matting. Let the curds set undisturbed for 5 minutes. Drain off the whey.

6. Add the salt and maintain the curds at 95°F (35°C) for another 30 minutes.

7. Gently ladle the curds into a basket cheese mold lined with cheesecloth. Press at 35 pounds of pressure for 6 hours, unwrapping, flipping, and rewrapping at 30-minute intervals.

8. Remove the cheese from the mold and place in a covered container. Store in the refrigerator for up to 1 week.

A CHEESE MAKER'S STORY

Ruth Appel
APPEL FARMS
Ferndale, Washington

APPEL FARMS, WHICH DATES BACK TO THE LATE 1960S, is located in Ferndale, Washington, two hours due north of Seattle and 20 minutes south of the Canadian border, with snowcapped Mount Baker as a backdrop. Visitors are welcome to stop by Monday through Saturday, from 8 A.M. to 5 P.M., and dine in the café — and, if they time it right, watch cheese being made through a viewing window.

"It all started with my father-in-law, Jack, who grew up in Holland with dreams of being a dairy farmer," says Ruth Appel, whose husband, John, currently runs the farm with his brother, Rich. When Jack moved to the United States and bought this land, he only made cheese as a hobby until he was approaching retirement and a German importer asked him to make quark. "No one was making or much less knew about quark back then, so Jack started making tiny batches in this 850-square-foot building, thinking the space was so big he would never be able to fill it."

That was pretty much the case for the next ten years, during which time Jack died (in 1999) and John took over the cheese making. "John grew up on this farm and wanted to carry on the legacy. The more he made cheese, the more he fell in love with it, particularly the research and development." The next product John made was feta, followed by paneer, at the request of the same German importer. "Now it's our biggest seller."

After adding on to that original building in several directions and spreading into two other buildings, they built a 12,500-square-foot creamery space in January 2017. "That may sound enormous, but I always tell new cheese makers that we started in a small refrigerator, 'just like you.'" She also advises them to grow the sales slowly, using distributors to help ease the load of fulfillment as well as having direct contact with nearby chefs (the farm's biggest proponents).

And despite venturing forth into other Dutch-style aged cheeses, Ruth still has a soft spot for the original Appel Farm cheese. "Quark is making a comeback. There's just no real substitute for it. Baked goods are so moist and delicious when made with quark."

QUARK

With a recorded history dating back to 3 BCE, quark is the oldest form of cheese in Europe. The taste is all its own but often described as what would happen if you crossed yogurt with fromage blanc, leaving you with a rich, creamy product that lacks yogurt's characteristic tang. It is superlatively versatile in sweet and savory cooking and just as delicious on its own for breakfast, topped with fresh fruit. It is especially popular in Germany.

Regular quark is made with whole milk; Magerquark (low-fat quark) is made with 1–2% milk. Sahnequark (cream quark) is made with additional cream for every gallon of whole milk. Yet another formula calls for 1 pint light cream or half-and-half with a gallon of whole milk. You can make whichever variety you prefer using this recipe.

1 gallon milk and/or cream (see headnote)

¼ teaspoon calcium chloride diluted in ¼ cup cool, nonchlorinated water

1 packet C20 fromage blanc starter culture*

2–3 tablespoons heavy cream (optional)

*You may substitute 1–3 drops of liquid rennet mixed with either 1 packet C21 buttermilk starter culture or ⅛ teaspoon C11 Flora Danica (see page 15).

Yield: 1–1½ pounds

1. Heat the milk to 86°F (30°C). Add the calcium chloride solution and stir well to combine. Sprinkle the starter over the surface of the milk, wait 2 minutes for the powder to rehydrate, then stir well.

2. Cover and let set at 72–86°F (22–30°C) for 12–24 hours, or until curds form.

3. Gently ladle the curds into a colander lined with butter muslin. Tie the corners of the muslin into a knot and place the bag of curds in the refrigerator to drain for 6–8 hours. You may put a container filled with water on top of the curd bag to press a bit and speed drainage.

4. Transfer the drained quark to a bowl. If the quark is too dry, combine the cream with the finished cheese. Cover the bowl and store in the refrigerator for up to 1 week.

30-Minute Mozzarella

This is a quick and easy way to make fresh mozzarella at home in less than 30 minutes if your milk is not ultrapasteurized. The protein in ultrapasteurized milk is denatured and it will leave you with a ricotta-like product rather than mozzarella. You may use skim milk in this recipe, but the yield will be lower and the cheese will be drier.

The curds are heated to a very high temperature to produce that famous mozzarella stretchiness. Have a pair of heavy rubber gloves and a wooden spoon handy for working the curd in hot water.

Variation

If all you can find is ultrapasteurized milk, a delicious alternative is to use dry milk powder and cream. Make 1 gallon of milk and let it sit overnight at room temperature. To make mozzarella, use 7 pints of the reconstituted milk with 1 pint of light cream or half-and-half. (Because of the ratio of cream to milk, the cream can be ultrapasteurized.)

1 teaspoon citric acid dissolved in 1 cup cool water

1 gallon pasteurized whole milk (not ultrapasteurized)

¼ teaspoon liquid rennet (or ¼ rennet tablet) diluted in ¼ cup cool, nonchlorinated water

1 teaspoon cheese salt (optional)

Yield: ¾–1 pound

Note: If using raw milk, in step 1 heat the milk to 88°F (31°C). Remove the pot from the heat and do not heat the curd again in step 3. It may take a few more minutes to set because of the high cream content, but you will be rewarded with a wonderfully rich and creamy mozzarella.

1. Pour the citric acid solution into the pot, add the milk, and stir well. Heat the milk to 90°F (32°C) then remove the pot from the stove. Some curdling will begin as the milk is heated, which is fine.

2. Add the rennet solution and stir slowly with an up-and-down motion for 30 seconds. Cover and let set for 5 minutes, or until a solid curd forms. Test it by gently pressing with the back of your hand on the curd near the side of the pot to see if it pulls away and shows some clear whey. If the curd is soft, wait another 5 minutes.

3. Cut the curd into 1-inch cubes. Return the pot to the stove and heat the curds to 105°F (41°C). Remove from the heat and stir slowly for 2–5 minutes. (The longer you stir, the firmer the final cheese will be.)

4. With a slotted spoon, scoop the curds into a colander set over a bowl. Gently press on the curd mass and fold it over on itself a few times to release more whey.

5. In another pot, heat a quart of water to 180°F (82°C).

6. Turn the curd mass onto a cutting board and cut it into ⅛-inch slices, similar to a brisket. Submerge the slices into the hot water and using a slotted spoon or your gloved hands, quickly work the pieces of curd by pressing them together and folding them over to heat evenly and for one mass. (You can do this step with half of the curds at a time for ease of handling.)

6. Once the pieces form a mass, begin to pull it out of the water with a spoon, letting the curd stretch down from the spoon and then folding it back on itself. Repeat several times until it becomes smooth and elastic. If it isn't stretching, check your water temperature and adjust if needed. If the curd becomes too cool, it will start to break rather than stretch. If this happens, dip it back into the hot water.

7. Once the curd stretches like taffy and develops a sheen, you can roll it into a ball, shape it into a braid, or leave it in long strands. Place the cheese into a bowl of 50°F (10°C) water for 5 minutes and then into a bowl of ice water for 15 minutes. Sprinkle with a pinch of salt, if using, and eat it up or store for up to 2 days in the refrigerator.

TROUBLESHOOTING

If the curds turn into the consistency of ricotta, try another brand of milk. It may have been pasteurized at too high a temperature. I find anything pasteurized over 171°F (77°C) will not work with this recipe.

A CHEESE MAKER'S STORY

Imran Saleh

Lahore, Pakistan

WHEN IMRAN SALEH, a self-described "small-scale businessman dealing in PVC and rubber pipes," went looking for quality cheese in Pakistan back in 2008, he soon realized he'd just have to make it himself. "I started experimenting with milk in my home kitchen, which was pretty much a disaster for the first few months, at least until I found a source for rennet." When his wife booted him out of the kitchen ("Enough is enough!"), he devised his own make room, building the small vats and other equipment himself.

In 2013, he finally made a batch of mozzarella that he was happy with, then proceeded to print some labels and see if he could sell some. Just eight months later, he opened his first commercial make room and trained four employees to produce his version of "dry" mozzarella, which he created as a better melting option to preshredded "pizza" mozzarella (pizza being a booming business in Pakistan).

By 2015, he was making more than 20 types of cheeses, seven days a week (while still running his other business). And now Imran has an even larger facility (air-conditioned, no less) with brand-new equipment and six employees who help process 264 gallons of milk each day and produce 40 different cheeses, including bocconcini, English cheddar, herbed Gouda, pepper jack, cream cheese, and halloumi. "I only use raw, unpasteurized milk from healthy, hormone-free cows and buffalo."

Imran credits passion as the reason he was able to turn his hobby into a profession. "During all these years I went through failures, disappointments, financial struggles, and countless other obstacles, including finding quality milk. But I never gave up, because this was something I loved to do. I kept moving forward, through both good and bad times. Eventually success knocked on my door." Today, Imran's company, Farmer's Cheese Making, supplies cheeses to restaurants in major cities throughout Pakistan and also to various foreign embassies. "I am not sure if I was the first person to produce artisan cheese in my country, but I am certainly the first to introduce such a large variety."

Imran also spends time training others and says a few of those people are selling in farmers' markets. "Hopefully it's going to be a revolutionary industry that will change the future of Pakistan."

QUESO
OAXACA,
page 108

30-MINUTE
MOZZARELLA, *page 92*

TRADITIONAL MOZZARELLA

Mozzarella was first made by the monks of San Lorenzo di Capua, Italy, from sheep's milk. When water buffalo were introduced to Naples in the sixteenth century, cheese makers began using the rich milk of those animals. If you can find that type of milk, you can use it to make the real deal. Otherwise, most American-made mozzarella is made with good old cow's milk.

This recipe takes longer to make and creates a more flavorful cheese than the 30-Minute Mozzarella (see page 92). Note that the curds are heated to a very high temperature (170°F/77°C) to produce the stretch for which mozzarella is prized. Have a pair of heavy rubber gloves and wooden spoons handy for working the mozzarella curd in hot water (step 6).

2 gallons milk

1 packet direct-set thermophilic starter culture

¼ teaspoon lipase powder dissolved in ¼ cup cool, nonchlorinated water and allowed to sit for 20 minutes, for stronger flavor (optional)

½ teaspoon liquid rennet (or ½ rennet tablet) diluted in ¼ cup cool, nonchlorinated water

Cheese salt

Yield: About 2 pounds

1. Heat the milk to 100°F (38°C). Sprinkle the starter over the surface of the milk, wait 2 minutes for the powder to rehydrate, then stir well. Add lipase solution, if desired, and stir. Cover and let ripen for 60 minutes.

2. Add the rennet solution and gently stir with an up-and-down motion for 30 seconds. Cover the pot and let set at 100°F (38°C) for 45 minutes, or until the milk forms a soft curd.

3. Cut the curds into 2-inch-square vertical columns and let sit for 5 minutes. Cut the curd into 1-inch cubes, both vertically and horizontally. Stir very gently for 2 minutes, then let the curds settle. Over the next hour, stir briefly every 3 minutes to keep the curds from matting.

4. Gently ladle the curds into a colander. Place the colander on a pot full of warm water, keeping the curds above the water. Cover and keep the curds at 96-100°F (36–38°C) for 2 hours.

5. To test the curd for stretching, cut off a ½-inch slice and place it into a cup of 180°F (82°C) water. Let sit for a minute, then try stretching a piece. If the curd stretches 2 to 3 times its original length without breaking, then you are ready to continue. If it does not, wait another 15 minutes and repeat the test with another piece of curd.

6. Place the curd on a cutting board propped up so that it can drain into a sink and cut the curds into ⅛-inch slices. Place a handful of slices into a container with 1–2 inches of 180°F (82°C) water. Using a wooden spoon, work the curd until it's melded together. With the spoon, pull the curd out of the water and watch it stretch like taffy. Fold the curd onto itself several times, then with your hands (gloves advisable), work the curds into a smooth and shiny ball. Work quickly; this is fun, but any overworking will result in a tough cheese.

7. Put the cheese into a bowl of 50°F (10°C) water for 15 minutes, then into ice water for 30 minutes so it will hold its shape. Salt to taste before serving. Store in the refrigerate for up to 2 days. It will keep for several months if placed in saturated brine (see page 46) and refrigerated.

FETA

Traditionally a highly salted Greek cheese made from sheep's milk; this version uses goat's milk, which naturally contains lipase enzymes and produces a stronger-flavored cheese. If you plan to make this recipe using cow's milk, you may want to add lipase to create a stronger flavor profile.

2 gallons goat's milk

¼ teaspoon lipase powder dissolved in ¼ cup nonchlorinated water and allowed to sit for 20 minutes (only if using cow's milk)

¼ teaspoon calcium chloride diluted in ¼ cup cool, nonchlorinated water

1 packet direct-set mesophilic starter culture or C21 butter-milk starter culture

¼ teaspoon liquid rennet (or ¼ rennet tablet) diluted in ¼ cup cool, nonchlorinated water

2 teaspoons cheese salt, plus more for step 9

½ gallon Saturated Brine (page 46)

Yield: 1 pound

1. Combine the milk and the lipase solution, if using cow's milk. Heat the milk to 93°F (34°C). Add the calcium chloride solution and stir well to combine. Sprinkle the starter over the surface of the milk, wait 2 minutes for the powder to rehydrate, then stir well. Cover and allow the milk to ripen for 1 hour.

2. Add the diluted rennet and stir gently with an up-and-down motion for 1 minute. Cover and let set for 40 minutes, or until the curds give a clean break.

3. Slowly cut the curd into ½-inch cubes. This should take about 5 minutes.

4. Gently stir the curds for 20–30 minutes, maintaining a temperature of 93°F (34°C). Then let the curds settle for 10 minutes.

5. Gently ladle the curds into a basket mold. Top them with a follower and add a 1-pound weight (a pint container full of water works well). Allow the curds to drain at 68–72°F (20–22°C) for 8 hours. During the first 2 hours, unmold the cheese, flip it, and return it to the mold frequently.

6. After 8 hours, cut the cheese into ½-pound pieces. Arrange the pieces on a draining mat and sprinkle with salt. Allow to drain for 6–12 hours.

7. Place the pieces in the saturated brine for 4 hours.

8. Remove the cheese from the brine and place on a draining mat. Cover loosely with a piece of butter muslin or cheesecloth and let dry at 48–56°F (9–13°C) for 1–3 days, turning the pieces several times a day.

9. Put the cheese in a lidded jar and cover completely with 8% brine (8 ounces of salt to 3 quarts of water). Age at 48–50°F (9–10°C) for up to 30 days. The cheese will keep in the solution for up to a year at 45–55°F (7–13°C). Younger cheese will have a milder flavor.

FETA

Bulgarian–Style Feta

Although the name "Feta" can only be officially used to refer to a certain cheese made in Greece, endless varieties of this salty white cheese are made throughout the world. This version, based on an age-old process from Bulgaria, relies on the yogurt culture bulgaricus, *appropriately named for the region of its origin.*

2 gallons milk

½ teaspoon calcium chloride diluted in ¼ cup cool, nonchlorinated water

4 ounces prepared Bulgarian yogurt

1 packet C21 buttermilk starter culture

¼ teaspoon liquid rennet (or ¼ rennet tablet) diluted in ¼ cup cool, nonchlorinated water

4 tablespoons cheese salt

1 gallon Saturated Brine (page 46)

Yield: 2 pounds

1. Slowly heat the milk to 88°F (31°C). Add the calcium chloride solution and stir well to combine. Add the yogurt. Sprinkle the buttermilk starter over the surface of the milk, wait 2 minutes for the powder to rehydrate, then stir well to incorporate both the culture and the yogurt. Cover the pot and let the milk ripen at 88°F for 30 minutes.

2. Add the diluted rennet and stir gently with an up-and-down motion for 30 seconds. Cover the pot and let set undisturbed for 60 minutes, or until the curd gives a clean break. Try to keep the milk warm, but it's okay if the temperature drops a few degrees during this time.

3. Cut the curd with long vertical cuts, each about 2 inches apart. Repeat at right angles to your first cuts, creating long columns of curd. Cover the pot and let the curds settle under the whey for 15 minutes.

4. Use a ladle to transfer the curds to a colander lined with butter muslin, gently cutting the columns into ½- to ¾-inch pieces with the ladle as you go. (The larger the curds, the moister the final cheese.) Fold the cloth over the curd and add a 4-pound weight to help the curd release whey. Let the curd drain until the flow of whey slows down, 60–90 minutes.

5. Open the cloth and cut the curd into 2- to 3-inch pieces. Gather the curd together in a tight ball, knotting the ends of the fabric together to create a smooth package with a knot at its top. Add the weight and let it rest for 60–90 minutes.

6. Open the cloth and repeat the process of breaking up the curd and reforming it into a snug bundle in the cloth. Flip the bundle over so it's resting on its knot, add the weight, and let it sit for 1–2 hours, longer for a drier cheese.

7. Open the cloth and break up the curd. Transfer the curd to two basket cheese molds set onto a draining tray. Pack the curds firmly into the molds. Put the molds in a warm place (70–80°F/21–27°C) and top each one with a 2-pound weight. Let the cheese rest and drain for 3–4 hours. Remove the weight and let rest for 8–10 hours.

8. Cut each cheese into four pieces and lay them out on a draining pan lined with butter muslin. Sprinkle the cheese on all sides with 1 tablespoon of the salt. Let it rest at 52–54°F (11–12°C) and 70–75% relative humidity for 12 hours. Rotate the cheese so that it shows a new face, rub all sides with another tablespoon of salt, and let rest for another 12 hours. Repeat the process two more times so the pieces are salted four times. Let the cheese rest for 4 days, flipping daily.

9. Pour the saturated brine into a glass jar with room to add the cheese. Submerge the cheese in the brine and age at 48–50°F (9–10°C) until the flavor of the cheese is developed to your liking (14–30 days).

10. Store the cheese in 8% brine (8 ounces of salt to 3 quarts of water) at 42–46°F (6–8°C) until consumed. If the finished product is too salty for your taste, soak it in milk for several hours before serving.

Halloumi

Originating in Cypress, halloumi is a good hot-weather cheese: The salt in the brine inhibits the growth of mold and unwanted bacteria, which usually thrive in temperate conditions. It is also ideal for cooking on the grill, turning smoky and blistered without melting.

2 gallons milk

¼ teaspoon calcium chloride diluted in ¼ cup cool, nonchlorinated water

1 packet direct-set mesophilic starter culture

½ teaspoon liquid rennet (or ½ rennet tablet) diluted in ½ cup cool, nonchlorinated water

¼ cup cheese salt

1 gallon Saturated Brine (page 46)

Yield: 2 pounds

1. Heat the milk to 86°F (30°C). Add the calcium chloride solution and stir well to combine. Sprinkle the starter culture over the surface of the milk, wait 2 minutes for the powder to rehydrate, then stir well.

2. Add the diluted rennet and stir gently with an up-and-down motion for 30 seconds. Cover and allow to set at 86°F (30°C) for 30–45 minutes, or until the curd gives a clean break.

3. Cut the curd into 1-inch cubes.

4. Increase the temperature 2 degrees every 5 minutes until the curds reach 104°F (40°C), stirring gently every 3 minutes to keep the curds from matting. Maintain the curds at 104°F for 20 minutes, stirring gently every 3 minutes.

5. Ladle the curds into a cheesecloth-lined colander. Drain the whey into a pot and reserve.

6. Pack the curds into a cheesecloth-lined mold and press for 1 hour at 30 pounds of pressure.

7. Remove the cheese from the mold and gently peel away the cheesecloth. Turn over the cheese, re-dress it, and press for 30 minutes at 50 pounds of pressure.

8. Remove the cheese from the mold and cut it into 4-inch-square blocks.

9. Bring the reserved whey to 176–194°F (80–90°C). Place the curd blocks in the whey and soak for 1 hour, at which time the cheese will have a texture similar to that of cooked chicken breast and will rise to the surface.

10. Ladle the curds into a colander and let cool for 20 minutes.

11. Sprinkle the curds with the salt and let cool for 2–4 hours.

12. Store the cheese in the brine at 45–55°F (7–13°C) for up to 60 days. The flavor increases with age, but the cheese may be eaten fresh at any time during the 60-day period.

CRESCENZA

This soft and spreadable cheese is a staple in Italian kitchens, and when it comes to expressing the natural flavor and aromatics of a good-quality milk, no other cheese does it better. Crescenza has a short shelf life due to its high moisture content, but it's easy to eat this delicious cheese quickly. This recipe is unique in that the salt is added before the culture. The salt slows the culture down, and this slower lactic acid production preserves the final high moisture level of the cheese. For this recipe, you will set up two mold sandwiches (see page 44), so you will need two square molds, four cheese boards, and six cheese mats.

1 gallon milk

¼ teaspoon calcium chloride diluted in ¼ cup cool, nonchlorinated water

1½ tablespoons cheese salt

2 ounces fresh yogurt prepared from Y1 Bulgarian yogurt culture

Just over ⅛ teaspoon liquid rennet (or ⅛ rennet tablet) diluted in ¼ cup cool, nonchlorinated water

Yield: 1 pound

Note: To give this cheese a longer shelf life, add a pinch of *Geotrichum candidum* along with the culture.

1. Slowly heat the milk to 100°F (38°C). Add the calcium chloride solution and the salt, and stir well to combine. Add the yogurt, stir well, and let the milk ripen at 100°F for 30 minutes.

2. Add the diluted rennet and stir gently with an up-and-down motion for 30 seconds. Cover and let set undisturbed at 100°F (38°C) for 60 minutes, or until the curd gives a clean break.

3. Cut the curd vertically into 2- to 3-inch columns. Do not stir; let rest for 30 minutes. The whey should be a nice clear yellow/green.

4. Cut the curds again, this time into ⅝- to ⅞-inch cubes. The larger the curd, the moister your final cheese will be. Try not to disturb the lines of your first cuts and be very gentle with the delicate curd. Stir just enough to keep the curds separated. Let the curds settle, gently stirring every 5 minutes, for 15 minutes.

5. Place two cheese boards in a draining pan. Put a mat and a Pont-Levesque mold on each board (the start of a mold sandwich). It's essential to keep the curds at 75–78°F (24–26°C) while they drain, so set the molds up either in a very warm room or a container in a water bath.

6. Using a colander in the pot to hold down the curds, scoop out whey until the liquid level is just at the level of the curds. Use a ladle to carefully transfer the curds and any remaining whey to the molds, dividing them equally.

7. Place a third draining mat over one of the molds and quickly turn the mold over. Repeat with the second mold, using the mat and board from the first mold. This will create a nice, smooth surface on the top of the curds as they continue to drain. Let the curds continue to drain for 3 hours, flipping the molds every hour. Then let the curds rest in the molds for 8–10 hours.

8. Remove the cheeses from the molds. Each cheese will be about 1½ inches thick and similar to firm pudding in texture. Transfer the cheese to a draining mat and store in a covered container in the refrigerator. Let it rest, turning it and draining any whey from the container, for 3–5 days. It will soften over time and become very spreadable. Eat within 10 days, bringing the cheese back up to room temperature before serving.

ROBIOLA

This simple, beautiful cheese is ideal for the home cheese maker. Robiola, which originally hails from outside Torino, Italy, is incredibly versatile. It requires very little active time and can be eaten fresh or aged. You can make this cheese with almost any milk — raw, pasteurized, milk from cow, goat, or ewe — and with any culture. But please, please, please, no low-fat or skim milk! This cheese depends on the natural flavor of milk with full butterfat. It can be eaten fresh after 2 or 3 days or allowed to develop a slight mold coat and aged for several weeks.

1 gallon milk

¼ teaspoon calcium chloride diluted in ¼ cup cool, nonchlorinated water

1 packet C21 buttermilk starter culture

4 drops liquid rennet diluted in ¼ cup cool, nonchlorinated water

1 gallon Saturated Brine (page 46)

Light Brine (page 46)

Yield: 1 pound

1. Slowly heat the milk to 72°F (22°C). Add the calcium chloride solution and stir well to combine. Sprinkle the starter over the surface of the milk, wait 2 minutes for the powder to rehydrate, then stir well. Cover the pot and let the milk ripen at 72°F for 4 hours.

2. Add the diluted rennet and stir gently with an up-and-down motion for 30 seconds. Cover and let set undisturbed until you see a thin layer of whey on top of the curds. This may happen in as little as 20–30 minutes, or it could take up to 8 hours.

3. Cut the curds vertically into 1½-inch columns. Let the curd rest for 5 minutes as the edges of the cuts firm up. Then make horizontal cuts, creating ⅜- to ½-inch-thick cubes. Stir the curds gently every 3 minutes for 10 minutes. Then let the curds settle to the bottom of the pot for 5 minutes. Set three basket cheese molds on a rack over a sink or large bowl and line each with butter muslin.

4. Remove enough whey from the pot so that the liquid is just to the curd level. Use a ladle to transfer the curd to the molds, letting the curd float in the whey as it drains. Pile the curds on the top of each mold. Fold the cloth over the top of each mold and flip each cheese in its mold. Let the cheeses drain for 10 minutes.

5. Unwrap the cheeses, flip them, and rewrap in the butter muslin. Let them continue to drain for 45–60 minutes. Then unwrap the cheeses and return them to the molds without their cloth wrapping. Let the cheese rest at room temperature or 72°F (22°C) for 12–18 hours.

6. Remove each cheese from its mold and submerge it in the saturated brine for 1 hour. Transfer the cheese to a cool space (55–65°F/13–18°C) to dry for 4–6 hours. Then age at 52–58°F (11–14°C) and 80–85% relative humidity, turning daily, for 4 days. If any mold appears on the surface of the cheese, wipe it away with light brine.

7. At this point, the cheese is ready to eat as a young, fresh cheese. Alternatively, you can age the cheese for up to 40 days, wiping the surface down with light brine every 2–5 days, or as needed.

Well, many's the long night I've dreamed of cheese — toasted, mostly . . .

ROBERT LOUIS STEVENSON III

QUESO OAXACA

Queso Oaxaca is a fresh stretched cheese from the Oaxaca region of Mexico, similar to fresh mozzarella. It's fun to pull apart and eat plain, but it's most often used in cooking and is by far the most popular cheese in Mexico for quesadillas.

2 gallons milk

¼ packet C201 direct-set thermophilic starter culture or 4 ounces fresh yogurt prepared from Y1 Bulgarian yogurt culture

¼ teaspoon liquid rennet (or ¼ rennet tablet) diluted in ¼ cup cool, nonchlorinated water

1 gallon Saturated Brine (page 46)

Cheese salt

Yield: 2 pounds

1. Slowly heat the milk to 94°F (34°C). Sprinkle the starter over the surface of the milk, wait 2 minutes for the powder to rehydrate, then stir well. (If you're using yogurt, simply stir it to eliminate any lumps and stir it into the milk.) Cover the pot and let the milk ripen at 94°F for 3 hours.

2. Add the diluted rennet and stir gently with an up-and-down motion for 30 seconds. Cover and let set undisturbed at 94°F (34°C) for 30 minutes, or until the curd gives a clean break.

3. Cut the curd vertically into ½- to ¾-inch-square columns. Let the curd rest for about 5 minutes, after which cut the curd horizontally into ½- to ¾-inch cubes, opting for a larger curd if you prefer a moister cheese.

4. Slowly heat the curds to 100°F (38°C) over the next 30 minutes, stirring once or twice. The final result should be firm curd, fully cooked through and resistant to the touch. Remove the pot from the heat and let the curd settle in the whey for 5 minutes. Using a colander in the pot to hold down the curds, scoop out the whey until the liquid level is just at the level of the curds. Transfer the curds to a colander or cheese mold lined with butter muslin.

5. Let the curd develop acidity while it drains at 95–100°F (35–38°C) for 2–4 hours, flipping the curd halfway through the draining process.

6. Toward the end of the draining time, heat a pot of water or whey to 180°F (82°C). Test the curd by cutting off a small piece and submerging it in the water for 5 minutes. If it stretches easily, it's had enough time to develop. Otherwise, give it 30 more minutes and test it again. If the curd stretches but breaks, repeat the test at 15-minute intervals, until it stretches without breaking.

7. Slice the curd mass into ¼-inch strips and transfer them to a bowl. Pour some of the hot water into the bowl, pouring it along the side of the bowl so it doesn't hit the curd directly. Use your hands or a spoon to keep the curd strips separate and evenly heated. Keep the curds separate until they begin to stretch on their own and the water cools to 145°F (63°C), about 5 minutes. Add more hot water if the water dips below 145°F. Once the curds are warm and stretchy, consolidate them back into a single mass.

8. Now stretch the curds. During this process, you'll want to either periodically dip your hands into ice water or wear heavy rubber gloves. Start by using a heavy wooden spoon to lift the large cheese mass out of the water. It should slowly stretch under its own weight. Let it stretch enough to fold double on itself, then submerge in the water and repeat 2 or 3 times. If the curd gets stiff, check the temperature of the water and add more hot water if needed.

9. Stretch the curd again by pulling it into ¾-inch-wide by ½-inch-thick bands like a rope. As you continue stretching and pulling, allow the ends to fall into a bowl of cold water. Leave it in the water for 10 minutes until chilled. Roll the bands into a ball, similar to rolling a ball of yarn.

10. Place the cheese in the saturated brine for 3–4 hours. The cheese will float, so sprinkle the top surface with salt, flip the cheese halfway through the soak time, and sprinkle the other side with salt. Eat the cheese fresh, or store in a covered container in the refrigerator for 7–10 days.

Small–Curd Cottage Cheese

This cheese has smaller curds and is coagulated by the action of starter culture bacteria instead of by rennet. It has a pleasantly sour taste and is delicious eaten alone or used in recipes calling for cottage cheese. If you stir the curds until they are firm and omit the addition of heavy cream, you will have "dry curd" cottage cheese, which contains less lactose— a good choice for people with sensitivities.

1 gallon milk

¼ teaspoon calcium chloride diluted in ¼ cup cool, nonchlorinated water

1 packet direct-set mesophilic starter culture

1–2 tablespoons heavy cream (optional)

Cheese salt (optional)

Yield: 1½ pounds

1. Heat the milk to 72°F (22°C). Add the calcium chloride solution and stir well to combine.

2. Sprinkle the starter over the surface of the milk, wait 2 minutes for the powder to rehydrate, then stir well. Cover and let set at 72°F (22°C) for 16–24 hours. The curd will be rather soft.

3. Gently cut the curd into ¼-inch cubes and let set for 15 minutes.

4. Increase the heat by 1 degree per minute until it reaches 100°F (38°C). Stir gently every few minutes to keep the curds from matting. Maintain the temperature at 100°F for 10 minutes, stirring once every 3 minutes or so to prevent the curds from matting.

5. Increase the temperature to 112°F (44°C) over a 15-minute period (a little less than 1 degree per minute).

6. Stirring gently every 3 minutes, maintain the temperature at 112°F (44°C) for 30 minutes, or until the curds are firm. Test for firmness by squeezing a curd particle between your thumb and forefinger. If it still has a custardlike consistency inside, it is not ready. Wait another 5–10 minutes and test again.

7. When the curds are sufficiently cooked, let them settle to the bottom of the pot for 5 minutes. Pour off the whey. Ladle the curds into a colander lined with cheesecloth. Tie the corners of the cheesecloth into a knot. Rinse the bag in a bowl of ice water to cool and place the bag in a colander to drain for 5 minutes. (For a less sour cottage cheese, dip the bag of curds several times into a bowl of cool water before rinsing in ice water and draining.)

8. Remove the cheese from the bag and place it in a bowl. Break up any pieces that have matted. If desired, add the heavy cream to produce a creamier texture. Add salt to taste, if using. Cover the bowl and store in the refrigerator for up to 1 week.

About Cottage Cheese

This soft, fresh, cooked-curd cheese originated in Eastern and Central Europe and was quite popular in colonial America, when the cheese was made in local cottages (hence the name). It is also called pot cheese, after the vessel it was usually made in at home.

In days gone by, cottage cheese was made from raw milk, which was poured into a pot and set in a spot that would stay fairly warm. In winter, that meant next to or on a cool corner of the cookstove. Within several days, because of the action of the bacteria present in unpasteurized milk, the lactic acid level would become so high that the milk protein would precipitate out into a soft, white curd.

This soft curd could then be treated in a variety of ways. It was often sliced, warmed to about 100°F (38°C) for several hours, then drained to produce a delightful sour cheese. Sometimes the curds were merely drained without cooking to produce a lactic-acid type of cheese. Other times, the curd was pressed after cooking to produce farmer's cheese. The cottage cheese that we know today is made by one of the two methods presented in this book.

LARGE–CURD COTTAGE CHEESE

Coagulated by the action of starter culture and a small amount of rennet, this cottage cheese has slightly larger, more toothsome curds than the previous recipe.

1 gallon milk

¼ teaspoon calcium chloride diluted in ¼ cup cool, nonchlorinated water

1 packet direct-set mesophilic starter culture

¼ teaspoon liquid rennet (or ¼ rennet tablet) diluted in ¼ cup cool, nonchlorinated water

Cheese salt (optional)

Yield: 1½ pounds

1. Heat the milk to 72°F (22°C). Add the calcium chloride solution and stir well to combine. Sprinkle the starter over the surface of the milk, wait 2 minutes for the powder to rehydrate, then stir well.

2. Add 1 tablespoon of the diluted rennet and mix gently with an up-and-down motion for 30 seconds. Cover and let set at 72°F (22°C) for 16–24 hours, or until the curd coagulates. The curd will be rather soft.

3. Cut the curd into ½-inch cubes. Allow to sit undisturbed for 10 minutes.

4. Increase the heat by 2 degrees every 5 minutes, until the temperature reaches 80°F (27°C). Stir gently every 3 minutes to prevent the curds from matting.

5. Increase the heat by 3 degrees every 5 minutes, until the temperature reaches 90°F (32°C), stirring gently every 3 minutes.

6. Increase the heat by 1 degree per minute, until the temperature reaches 110°F (43°C), stirring gently every 3 minutes.

7. Maintain the temperature at 110°F (43°C) and stir gently every 3 minutes for 20 minutes, or until the curds are sufficiently cooked and no longer have a custardlike interior.

8. Continue the recipe as for Small-Curd Cottage Cheese, starting with step 7 on page 111.

A CHEESE MAKER'S STORY

Muwonge Baker
SEASONS DAIRY
Baale, Uganda

MUWONGE BAKER HAS BEEN MAKING AND SELLING CHEESE in Uganda for more than 16 years, an undertaking that has required equal parts persistence and passion. "My father, a dairy farmer, convinced a local producer to teach me how to make cheese over a six-month apprenticeship. I was interested in the whole process from the very first day, and that interest has continued to grow. I was happy to see the milk change to something right before my eyes."

Training completed, Baker returned to his home village, in the "dairy corridor" about 75 miles from the capital city of Kampala, to start his own cheese-making venture. "I started with about 50 liters (13 gallons) of milk to make about 5 kilograms (11 pounds) of cheddar cheese in the traditional way that I had been trained." Unfortunately, the cheese wouldn't set, and it took much trial and error (and more help from his former teacher) before he found his way. Baker also had a tough time convincing local farmers to bring him milk only three days a week because he used the other days to travel back and forth from Kenya where the cheese was packaged and distributed — all by him. Finding a market was yet another challenge, as he was, and still is, one of only a few cheese makers in Uganda.

Looking back, Baker says he has benefited from the help of private and public agencies — and even the president of Uganda — that took an interest in his operation and provided support. Case in point: Land O'Lakes, a nonprofit international agricultural development organization, helped Baker update his techniques. "After that extra training, the quality of my cheddar improved, the sales went up, and I managed to save some money to buy advanced machines from Kenya. I also hired a professional cheese maker from an old Kenyan cheese factory to work for me, and he is now the manager of the plant." He has since hired 15 other employees and branched out to make mozzarella, Gouda, feta, and paneer.

Says Baker, "I'm proud that my creamery has changed the lives of many people in my home village, where the main economic activity is dairy farming. I built the factory to provide a direct market for their milk. That is what drives me to keep working and striving despite setbacks — that and my undying passion for the art of cheese making."

SOFT AND SEMI-SOFT RIPENED CHEESES

Ripened cheeses owe much of their flavor and texture to specific bacteria and molds that are added to the milk and/or applied to the surface of the cheese. White molds and red bacteria grow on the surface of cheeses, and blue molds can grow both on the surface and throughout the cheese. Bacteria-ripened cheeses, such as Limburger, originated in Belgium and Germany; mold-ripened cheeses, including Camembert and blue cheese, were first produced in France.

In places where bacteria- and mold-ripened cheeses have been produced for hundreds of years, the specific bacteria and molds are in the air and on the shelves of the caves and rooms in which the cheeses are aged, automatically inoculating the freshly made cheeses. In those cases, new bacteria and molds are added only periodically to the milk or the cheese. Beginning cheese makers, however, need to purchase bacteria and molds from cheese-making supply houses.

CAMEMBERT, *page 122*

115

If you are making only one variety of cheese within this group, you don't have to worry about cross contamination. But if you are making several types of cheese, I recommend aging each type of bacteria- and mold-ripened cheese in a separate storage area and away from any nonbacterial- and non-mold–ripened cheeses, because these microorganisms can rapidly spread in your kitchen.

The following pages provide an overview of the basic steps involved in the production of bacteria- and mold-ripened cheeses; for a more thorough review, see chapter 3.

RIPENING

For most soft mold-ripened cheeses, a fairly large amount of starter is added, and the ripening time can be rather lengthy, as a significant increase in acidity is required.

RENNETING

A comparatively small amount of rennet is used to set most mold-ripened cheeses. It may take 1 hour or more for the milk to coagulate, and the curd is slightly softer than that produced in hard cheeses.

CUTTING THE CURD

The curd is usually cut into ½-inch cubes, or sometimes into thin slices, and gently ladled into a cheese mold.

COOKING

The cooking temperature is quite low, usually around 90°F (32°C). Some recipes omit this step.

MOLDING AND DRAINING

The curds are placed into cheese molds, most of which have traditional shapes and sizes.

Many cheeses are placed into molds that are sandwiched between two boards and two mats. (See also page 44.) In this "mold sandwich," the mold is placed on a cheese mat, which rests on a cheese board that is situated in a sink or pan to facilitate drainage. The mold is filled to the top with curd. Drainage begins immediately.

A second cheese mat is placed on top of the mold and a wooden cheese board is placed on the mat. Because these are unpressed cheeses, the curd sinks under its own weight. After a period of draining, the cheese is flipped over. The cheese mat that is now on the top is gently peeled off the cheese to allow the cheese to fall to the bottom of the mold. The turning process continues as the cheese drains for 24 to 36 hours.

SALTING

Once the cheese has sufficiently drained and assumed the shape of the mold, it is removed from the mold and salted on all surfaces. The salt helps retard the growth of unwanted organisms but does not interfere with the desired mold growth. Salting also draws out some of the moisture and is helpful in preserving the cheese.

APPLYING BACTERIA AND MOLDS

Carefully following a recipe's recommendations for proper aging, temperature, and humidity is of the utmost importance: Optimal growth of desirable bacteria and molds is dependent on very specific conditions.

PENICILLIUM CANDIDUM (WHITE MOLD)

This mold is used to age Camembert, Brie, and related cheeses. It may be added directly to the milk (enhancing early growth) or sprayed onto the surface of the cheese. For small-batch home use I recommend adding the mold directly to the milk; besides being the more economical method, it also makes it easier to control moisture levels.

GEOTRICHUM CANDIDUM (WHITE MOLD)

This mold is added to cheese in the same way as *Penicillium candidum* and will prepare the surface for a better growth of that mold.

PENICILLIUM ROQUEFORTI (BLUE MOLD)

This mold, which is cultivated on rye bread, is used for making blue cheeses. The liquid is added to the milk before renneting and grows throughout the cheese. Another technique is to pierce a cheese with a needle to create air holes. The mold, which requires oxygen, can then penetrate the surface and continue growing within the interior.

BREVIBACTERIUM LINENS (RED BACTERIA)

This reddish brown bacterium is sprayed onto cheese, then smeared on the surface by hand.

AGING

Bacteria- and mold-ripened cheeses need to be aged in a rather cool, moist setting in order to allow the bacterium or mold to grow. The temperature is usually started at 55°F (13°C), and lowered to 45°F (7°C) once mold forms. The relative humidity is 85 to 95 percent. Creating this environment in the home is a challenge. A secondhand refrigerator may be useful. A camping or dorm-size refrigerator is great because it's small, and the more cheese you put in it, the more humidity will be present. Place a tray of water on the bottom of the refrigerator for moisture and keep the setting at 45°F (7°C). Age the cheeses on mats, so that air can circulate freely underneath them.

If an extra refrigerator will not fit into your budget, you may try to age these cheeses in your regular refrigerator. (The drawback to using the household refrigerator is that at such low temperatures, bacteria and mold will not have the ideal conditions for reproducing and it may be far too cold for mold to develop properly.) If you do attempt this method, place the cheese on a cheese mat and put it in a closed container in which you have placed a bowl of water for moisture.

White mold–ripened cheese develops a furry, thick white coat within 2 weeks. Then the cheese is wrapped in cheese wrap and aged 6 weeks longer. Blue mold–ripened cheese develops mold within 10 days. At that time, the cheese must be pierced with holes so that oxygen will reach the interior and aid mold growth. In red bacteria–ripened cheese, bacteria will appear in 2 weeks and look like a reddish brown smear on the surface of the cheese.

CHEESE CAVES

No one knows for sure how someone figured out that molds and other organisms could improve the flavor of cheese. We'll just give credit to serendipity. Those who make Roquefort cheese tell the story of a young shepherd who left his lunch in a cave near the village of Roquefort; when he returned 2 weeks later, the bread had crumbled away and his cheese had veins of green growing through it. The hungry boy nibbled on the cheese and voilà! He shared his discovery with the villagers, and a new cheese was born.

You can still visit the Roquefort caves in France, where wheels and wheels of cheese are aged on racks in arched caverns, as well as cheese caves in England (where cheese has been ripened in the caves around Cheddar since the twelfth century), Italy, Spain, Canada, the United States, and other countries. In Sogliano al Rubicone, Italy, the opening of the Fossa caves on St. Catherine's Day in late November is the occasion of an annual festival and a tasting party for the cheeses aged in the caves since August.

Because not all cheese-making areas are blessed with the proper geological conditions that create caves, cheese makers have learned to build their own to mimic the steady temperatures and high humidity of natural caves. Natural stone, cast concrete, large cement culverts, and even straw bales have been used to build caves. If you tour cheese caves, you can see cheeses in various stages of finishing. And if you're lucky, you may be offered a taste.

Cheese ripening in a cave, Auvergne, France

WHITE MOLD— RIPENED CHEESES

These soft white cheeses are rich, creamy, and delectable, if not downright indulgent. In making them, the curds are typically not cooked or pressed. Instead, they are inoculated with mold spores to produce a rich white "bloom" on the surface of the cheese (my recommendation for home cheese makers).

CAMEMBERT,
page 122

PETIT BRIE

Brie is a truly remarkable cheese that is worth the care needed in making it. Due to its low acidity and high moisture content, it is susceptible to issues in handling, curing, and salting. It ages at a warmer temperature than other mold-ripened cheeses, and ripening must be done under optimum conditions.

The better Bries have a reddish brown mottling caused by Brevibacterium linens *(red bacteria) and a pale yellow interior, which when truly ripe is almost like custard. Traditionally, red bacteria are present in the aging room; however, if you are making this at home for the first time, you may add a small amount of the bacterium to the milk during the cheese-making process.*

2 quarts milk

¼ teaspoon calcium chloride diluted in ¼ cup cool, non-chlorinated water

⅛ teaspoon Flora Danica* starter culture

1⁄16 teaspoon *Penicillium candidum* (white mold)

1⁄16 teaspoon *Geotrichum candidum* (white mold)

1⁄32 *Brevibacterium linens* (red bacteria)

8 drops liquid rennet diluted in ¼ cup cool, nonchlorinated water

Cheese salt

*Flora Danica may be used as either a direct-set or a reculturable starter (see page 15).

Yield: 12 ounces

TROUBLESHOOTING

As with all recipes using small amounts of rennet, you may need to adjust the amount. If the cheese is too hard, use less next time; if the curds fail to set properly, use a little more next time.

Insufficient draining can result in the formation of a blue mold and/or a too-runny interior. If you see a black mold, try adding more salt or lowering the humidity. Dark, dry areas often result from excessive drying, as does the absence of the desirable white mold on the surface, which could also happen when dried at too-cold temperatures.

1. Heat the milk to 86°F (30°C) (84°F/29°C if using goat's milk). Add the calcium chloride solution and stir well to combine. Sprinkle the starter, *P. candidum, G. candidum,* and *B. linens* over the surface of the milk,

wait 2 minutes for the powders to rehydrate, then stir well. Cover and allow the milk to ripen for 15 minutes.

2. Add the diluted rennet and stir gently with an up-and-down motion for 30 seconds. Cover and let set undisturbed at 84–86°F (29–30°C) for about 12 hours. It will coagulate slowly.

3. Cut the curd into ½-inch cubes.

4. In a tray, place a mat on a cheese board and a Camembert mold on the mat (the start of a mold sandwich; see page 44).

5. Carefully pour off the whey. Gently ladle the curds into the mold. Place a mat on top of the mold and another cheese board on top of the mat (the finished mold sandwich).

6. Drain at a temperature of 68–70°F (20–21°C) for 12 hours. When the curds have settled to half their original volume, very carefully lift the mold and the cheese boards with both hands and quickly flip over the cheese. If the curd is stuck to the mat, gently peel it back and the cheese will drop down in the mold. You do not want to tear the surface of the cheese, which will happen if you lift the mat too quickly. Continue with draining.

7. When the cheese is firm enough to retain its shape (usually the next day), gently rub salt on the top and sides of the cheese. After a few hours, turn and gently rub salt on the bottom.

8. Allow the cheese to ripen at 58–64°F (14–18°C) and 55–65% relative humidity for about 1 week. When the white mold appears, age the cheese at 52–54°F (11–12°C) and 90–95% relative humidity for 6–8 weeks, or until the cheese is soft to the touch. Alternatively, after the white mold appears, wrap the cheese in cheese wrap and age it at 42°F (6°C). The wrap helps keep the moisture levels balanced, eliminating the need for high humidity.

CAMEMBERT

Camembert was developed in the French village of the same name in 1791 and soon came to be one of the most prized cheeses in the world. This deliciously runny cheese requires exacting conditions during aging, but you will be rewarded for your efforts.

One-half gallon of milk will make enough curd to fill a standard Camembert mold. Once the cheese has been opened, it stops aging; therefore, make several at a time so you can periodically test for ripeness.

2 gallons milk

¼ teaspoon calcium chloride diluted in ¼ cup cool, nonchlorinated water

¼ teaspoon Flora Danica* starter culture or 1 packet buttermilk starter culture

⅛ teaspoon *Penicillium candidum* (white mold)

¹⁄₃₂ teaspoon *Geotrichum candidum* (white mold)

¼ teaspoon liquid rennet (or ¼ rennet tablet) diluted in ¼ cup cool, nonchlorinated water

4 teaspoons cheese salt

*Flora Danica may be used as either a direct-set or a reculturable starter (see page 15).

Yield: 2 pounds

1. Heat the milk to 90°F (32°C). Add the calcium chloride solution and stir well to combine. Sprinkle the starter, *P. candidum*, and *G. candidum* over the surface of the milk, wait 2 minutes for the powder to rehydrate, and stir well. Cover and let the milk ripen at 90°F for 90 minutes.

2. Add the diluted rennet and gently stir with an up-and-down motion for 30 seconds. Cover and let set at 90°F (32°C) for 60 minutes, or until the curd gives a clean break.

3. Start four mold sandwiches with four mats, four cheese boards, and four Camembert molds (see page 44).

4. Using one hand to steady the mold, gently ladle thin slices of the curd into the molds, round-robin style, until they are heaped full. If there are more curds left, rather than using another mold, let the curds in the molds settle and then add more to each mold, until all the curd is used. Place a second mat and board on top of each mold (the finished mold sandwich).

5. Let drain for 1 hour. Very carefully lift each mold with its cheese boards and quickly flip them over. Make sure the cheese is not sticking to the top mat by gently peeling the mat from the top of the mold. You do not want to tear the surface of the cheese, which will happen if you lift the mat too quickly. Turn at hourly intervals, until the cheeses are about one-third of their original height and have shrunk away from the sides of the molds.

6. When they have sufficiently drained, remove the cheeses from the molds and sprinkle ½ teaspoon salt over the top and sides of each cheese and place back in the molds salted side up. After 5 hours, flip the cheeses, remove them from the molds, salt the other side and return them to their molds for 6 hours.

7. Unmold the cheeses and place them on the cheese mats to dry at 58–65°F (14–18°C) and 70% relative humidity for 3–6 days, turning them twice daily.

8. Once the surface has dried, move the cheeses to an aging space at 52–56°F (11–13°C) and 92–95% relative humidity. Turn once or twice daily. When the first whiskers of mold appear on the surface (after 9–14 days), move the cheeses to a cooler space, at 45°F (7°C). During the next 10–14 days, a profuse white mold will develop. Wrap the cheeses in cheese wrap and store at 45°F for 3–6 weeks. The cheeses are ready to eat when they are soft to the touch at room temperature.

TROUBLESHOOTING

If blue mold develops on the cheese, your curing room is too humid or you are allowing too much moisture to remain in the cheese during its production. Reduce the humidity in the curing room and thoroughly clean and disinfect all shelves to eliminate blue mold.

If you see a black furry mold starting to develop, dab it carefully with salt to remove it and prevent it from spreading. That mold is called *poil de chat* (cat's hair), and you do not want it on your cheese!

FRENCH–STYLE COULOMMIERS

A soft, white mold–ripened cheese, coulommiers is a runny cheese in the same family as Brie and Camembert with a long tradition in France, where it was created. The customary mold for coulommiers is a two-piece affair consisting of two stainless-steel hoops, one of which fits inside the other. Coulommiers is French for "columns," and that is what the mold looks like. As an alternative, you may use a Camembert mold.

1 gallon milk

¼ teaspoon calcium chloride diluted in ¼ cup cool, nonchlorinated water

¼ teaspoon direct-set Flora Danica* starter culture

¹⁄₁₆ teaspoon *Penicillium candidum* (white mold)

⅛ teaspoon liquid rennet diluted in 2 tablespoons cool, nonchlorinated water

Cheese salt

*Flora Danica may be used as either a direct-set or a reculturable starter (see page 15).

Yield: 1½ pounds

1. Heat the milk to 90°F (32°C). Add the calcium chloride solution and stir well to combine. Sprinkle the starter and the *P. candidum* over the surface of the milk, wait 2 minutes for the powders to rehydrate, then stir well. Cover and allow the milk to ripen for 20 minutes.

2. Add the diluted rennet and gently stir with an up-and-down motion for 30 seconds. Cover and let set at 90°F (32°C) for 45 minutes.

3. Set up two Camembert molds, two cheese mats, and two boards, the start of a mold sandwich (or put both molds on one large mat and board). Set them in an area where they can drain freely.

4. Using one hand to steady the molds, gently ladle thin slices of the curd into the molds, round-robin style, until heaped full. As the curds settle in the molds, continue filling until all curds are used. Be careful not to tip the molds or all the curd will rush out the bottom. Place a second cheese mat and board on each of the filled molds (the finished mold sandwich). Let the cheese set at 72°F (22°C) for 6–9 hours.

5. When the curd has drained halfway down into the molds, very carefully lift the mold and cheese boards with both hands and quickly flip over the cheese. Set down the mold and gently remove the top mat, making sure not to tear the cheese, which may be sticking to it slightly. Replace the cheese mat and board.

6. Allow the cheese to drain for up to 2 days, flipping it several times a day and carefully peeling away the top mat each time. The cheese is done when it stands 1–1½ inches in height and has pulled slightly away from the sides of the mold.

7. Remove the cheese from the mold and lightly salt it on all surfaces.

8. Age the cheese at 45°F (7°C) and a relative humidity of 85–95% for 5 days. After 5 days, small whiskers of white mold will appear on the surface of the cheese.

9. Turn over the cheese and age it 9 days longer. At that time, the cheese will be covered with a thick growth of white mold.

10. Wrap it in cheese wrap and age it for 4–6 weeks at 45°F (7°C).

TROUBLESHOOTING

If you see a black furry mold starting to develop, dab it carefully with salt to remove it and prevent it from spreading. This mold is called *poil de chat* (hair of the cat), and you do not want it on your cheese!

ENGLISH–STYLE COULOMMIERS

This soft, spreadable take on coulommiers is meant to be consumed fresh. To make it, simply follow steps 1 through 6 of the French-Style Coulommiers recipe, leaving out the *P. candidum*. Once you remove it from the mold, lightly salt all over and eat right away or wrap in cheese wrap and age for 1 to 2 weeks.

For a delicious variation, mix in fresh herbs, spices, and other seasonings in step 4: As you ladle the curd slices into the mold, sprinkle the seasonings between them. I like to add chopped chives, a pinch of freshly ground black pepper, and a dash of paprika.

LIMBURGER,
page 131

These cheeses are known for their strong smell, which is caused by red bacteria, and their mild taste, due to the washed-rind method of production. The degree of flavor development varies with the length of aging, which will depend on your personal preferences.

MUENSTER,
page 134

Brick Cheese

Developed in the United States, brick cheese has a characteristic rectangular shape that is about 10 by 5 by 3 inches. A reddish brown bacterium (Brevibacterium linens) grows on the surface and gives the cheese its mild, nutty flavor. It is affectionately known as "the married man's Limburger" because it is not as strong as the real Limburger.

It can be made in the traditional rectangular mold or in any mold that allows adequate drainage. Because the flavor is in large measure a result of the bacterial growth, the cheese cannot be thicker than the previously stated dimensions or the bacterial enzymes and flavor components will not penetrate the interior of the cheese.

2 gallons milk

½ teaspoon calcium chloride diluted in ¼ cup cool, nonchlorinated water

1 packet direct-set mesophilic starter culture

1⁄16 teaspoon *Brevibacterium linens* (red bacteria)

½ teaspoon liquid rennet (or ½ rennet tablet) diluted in ¼ cup cool, nonchlorinated water

1 gallon Saturated Brine (page 46)

Cheese salt

1 gallon Light Brine (page 46)

Yield: 2 pounds

1. Heat the milk to 88°F (31°C). Add the calcium chloride solution and stir well to distribute. Sprinkle the starter culture and *B. linens* over the surface of the milk, wait 2 minutes for the powders to rehydrate, then stir well. Cover and allow the milk to ripen for 10 minutes.

2. Add the diluted rennet and stir gently with an up-and-down motion for 30 seconds. Cover and let set at 88°F (31°C) for 30 minutes, or until the curd gives a clean break.

3. Cut the curd vertically into 1-inch columns and let rest for 2 or 3 minutes. Then cut the curds into ½-inch pieces with a wire whisk and rest again for 5 minutes, stirring gently every minute. The curd at this time is quite soft, so careful stirring is important until it firms up. Stir very gently over the next 10 minutes to prevent the curd from matting.

4. Over the next 60 minutes, heat the curds, increasing the temperature by 1 degree every 5 minutes, to a final temp of 100°F (38°C). Stir gently every 3 minutes to prevent the curds from matting.

5. Allow the curds to settle for 5 minutes.

6. Using a colander in the pot to hold down the curds, remove ½ gallon of whey and replace it with 65°F (18°C) water. The temperature will drop to 85°F (29°C). Continue stirring gently every 3 minutes, until curds have a moderate resistance when pressed between your fingers and are firm enough to mold, 15–30 minutes.

7. Let the curd set, undisturbed, for 5 minutes, then drain off the whey to the level of the curd.

8. Gently ladle the curds into a 2-pound rectangular mold. Allow to drain for 15 minutes with no weight. Turn the cheese in the mold and drain for 15 minutes with no weight. Turn the cheese in the mold again and apply 2 pounds of weight for 2–3 hours. Remove the weight, flip the cheese in the mold, and let sit in the mold with no weight overnight.

A small garden, figs, a little cheese, and, along with this, three or four good friends — such was luxury to Epicurus.

FRIEDRICH NIETZSCHE

Recipe continues on next page

9. Remove the cheese from the mold and soak in the saturated brine for 5 hours at 72°F (23°C), sprinkling a pinch of salt on the exposed surface of the cheese. Flip halfway through the brining process and salt the exposed surface.

10. Remove the cheese from the brine and pat dry with sanitized butter muslin.

11. Keep the cheese at 68°F (20°C) with a relative humidity of 90%. Wash and rub the cheese daily with a light brine for 7–10 days. By then it should have developed a good bacterial growth on the surface. At this point, it can be turned daily for the next 7–10 days without washing. It can then be wrapped with a 2-ply washed-rind paper and stored at 42°F (6°C) turning it twice weekly, until ready, 1–2 months depending on pungency desired.

LIMBURGER

Limburger is a soft, surface-ripened cheese with a strong flavor and aroma. It was first produced in Liège, Belgium, and is named for the province of Limbourg. Limburger is made from pasteurized milk in a process similar to that of brick cheese but is softer bodied and has a higher moisture content. It is rectangular like brick cheese, but somewhat smaller, which accounts, in part, for the more pronounced flavor: There is more surface area for bacterial growth in relation to volume.

2 gallons milk

¼ teaspoon calcium chloride diluted with ¼ cup cool, nonchlorinated water

1 packet buttermilk starter culture or ¼ teaspoon MM100 starter culture

1/32 teaspoon *Brevibacterium linens* (red bacteria)

1/16 teaspoon *Geotrichum candidum* (white mold)

½ teaspoon liquid rennet (or ½ rennet tablet) diluted in ¼ cup cool, nonchlorinated water

1 gallon Saturated Brine (page 46)

Cheese salt

Light Brine (page 46)

1/16 teaspoon *B. linens* (optional)

Yield: 2 pounds

1. Heat the milk to 88–90°F (31–32°C). Add the calcium chloride solution and stir well to combine. Sprinkle the starter, the *B. linens*, and the *G. candidum* over the surface of the milk, wait 2 minutes for the powders to rehydrate, then stir well. Cover and allow the milk to ripen for 30 minutes.

2. Add the diluted rennet and stir gently with an up-and-down motion for 30 seconds. Cover and let set at 90°F (32°C) for 45–60 minutes, or until the curd gives a clean break.

3. Being very gentle with this soft curd, cut into ½- to ⅝-inch cubes. Limburger curds have a jellylike consistency and are softer than the curds of brick cheese. Stir gently every 3 minutes for 15 minutes to keep the curds from matting.

4. Increase the temperature of the curds very slowly over 5–10 minutes to 92°F (33°C).

Recipe continues on next page

5. Stir the curds for 15–20 minutes, maintaining the temperature at 92°F (33°C). To check the curd for the right consistency, press it lightly in your hand. It should just barely stick together with a little thumb pressure.

6. Let the curds settle under the whey for 5 minutes. Using a colander in the pot to hold down the curds, remove the whey until the liquid level is at the curd level.

7. Set up four mold sandwiches using Camembert molds (page 44). Gently ladle the curds and whey into the molds. Once the molds are filled, flip the mold sandwiches. Flip the molds every 20 minutes for the first hour, then every hour for the next 3 hours. Continue draining for 12 hours, keeping the temperature at 72–78°F (22–26°C).

8. Remove the cheeses from the molds. They will be roughly 1½ inches tall. Place the cheeses in the saturated brine, sprinkle the top surface of the cheeses with salt, and let rest for 45 minutes. Flip the cheeses, sprinkle salt on the top surfaces, and let rest for another 45 minutes. Remove the cheeses from the brine and wipe the surface dry.

9. Age at 58–62°F (14–17°C) and 95% relative humidity, flipping daily, for 3–4 days.

10. Prepare a light brine solution. Wash the surface of the cheese with a cloth or soft brush, removing the slippery surface for a milder cheese. If you prefer a more aromatic cheese, add 1/16 teaspoon of *B. linens* to the light brine and only wash the surface lightly. Return to the aging space. Repeat the washing every other day for 10 days.

11. Continue aging the cheese at 40–50°F (4–10°C) and 95% relative humidity. After 2–3 months the cheese becomes soft and spreadable. After 3 months the cheese becomes strong and pungent. The surface needs to remain moist but not "gooey" during aging. If it begins to dry out, wipe with a damp cloth or lightly spray with a fine mist.

If I had a son who was ready to marry, I would tell him, "Beware of girls who don't like wine, truffles, cheese, or music."

COLETTE

NETTLE MEADOW, a farmstead goat's- and sheep's-milk creamery, is owned by Sheila Flanagan and Lorraine Lambiase, former California-based life partners and happy home cheese makers who dreamed of leaving their jobs in the legal field for full-time cheese making while searching for property in upstate New York, to be near their families. Lo and behold they found one, moved across the country, and took up where the previous owner left off, tweaking the original Kunik cheese, a triple-cream wheel, to give it more shelf life. "We were lucky enough to get picked up pretty early on by Murray's and Saxelby in New York City, which gave us some street cred and allowed us to branch out around the country."

In the 12 years since, while the number of cheeses they produce has grown, their staff and facilities haven't. "We're only 15 to 20 people working in half of a barn. We have a very small footprint and there is no mechanization helping us with the product. We walk the cheese down to the aging room, which is a butter cellar from the 1800s. It's all ladled out by hand into individual molds. The ongoing challenge is that people like things to be the same size, the same flavor profile, the same consistency when they buy it in June as in December."

All that effort adds to the price of the cheese, as does the cost of complying with food-safety regulations. Plus, as Sheila puts it, she wants to make sure their employees earn enough to buy groceries for their families and have some quality of life. "There is a reason artisan cheese costs what it does."

On top of running the creamery, the animal-welfare advocates created a sanctuary for their own elderly animals at the property (and, later, another one in New Jersey). When word got around — rather quickly, it seems — people started dropping off retired or rescued barnyard animals, including llamas, horses, pigs, turkeys, ducks, and donkeys (plus a bunch of kittens and cats). "We say 10 to 30 cents of each purchase goes to the quality of life of the animals, whether it be the new kids or the 35 retired goats who are out to pasture. People are buying into the particular way of food production that is meaningful to us."

A CHEESE MAKER'S STORY

Sheila Flanagan and Lorraine Lambiase
NETTLE MEADOW
Warrensburg, New York

MUENSTER

Muenster is an American version of Munster-style cheese, which originated in the Alsace region of France and dates back to the Middle Ages. Although traditionally made with whole cow's milk, it is also wonderful made with goat's milk. This semi-soft cheese develops deep flavor in a relatively short amount of time.

2 gallons milk

½ teaspoon calcium chloride diluted in ¼ cup cool, nonchlorinated water

1 packet direct-set mesophilic starter culture

1/16 teaspoon *Brevibacterium linens* (red bacteria)

1/64 teaspoon *Geotrichum candidum*

½ teaspoon liquid rennet (or ½ rennet tablet) diluted in ¼ cup cool, nonchlorinated water

Cheese salt

Light Brine (page 46)

1. Heat the milk to 90°F (32°C). Add the calcium chloride solution and stir well to combine. Sprinkle the starter, *B. linens*, and *G. candidum* over the surface of the milk, wait 2 minutes for the powders to rehydrate, then stir well. Cover and let ripen at 90°F (88°F/31°C for goat's milk) for 60 minutes.

2. Heat the milk to 95°F (35°C). Add the diluted rennet and stir gently with an up-and-down motion for 30 seconds. Cover and let set for 45–60 minutes, or until the curd gives a clean break.

3. Cut the curd into ½-inch cubes and stir very gently for 30 minutes.

4. Line two or three small Tomme molds with butter muslin. Gently ladle the curds into the molds. Place the follower on the molds and stack them on top of each other for 30 minutes at 75–80°F (24–27°C).

5. Unwrap the cheeses, flip, rewrap, and return to the molds. Restack them, switching the order from top to bottom. Let them drain at 75–80°F (24–27°C) for 30 minutes.

6. Transfer the molds to a space at 65°F (18°C) for 18–24 hours. Flip and restack the cheeses, switching the stacking order, every 5 or 6 hours. If checking the pH, it should be 5.2–5.3.

Yield: 2 pounds

7. Unwrap the cheeses and apply 1 teaspoon of salt to each cheese. Rub the salt over the entire surface and let sit at 65°F (18°C) and 80% relative humidity for 8 hours. Repeat the process with another teaspoon of salt per cheese and let sit for 2 days.

8. Age the cheeses at 57°F (14°C) and 95–98% relative humidity for 2 weeks, flipping daily. Wash the cheeses 3 times a week with a light brine. After 2 weeks, age at 43–46°F (6–8°C) and 95% relative humidity for 4–6 weeks, turning every other day. At 3–5 weeks, you will start to see a yellow-orange surface developing. This is the *B. linens* giving the cheese the desired aromatic flavor.

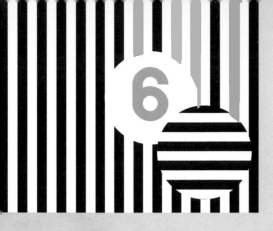

HARD AND SEMI-HARD CHEESES

aking hard cheese transforms milk protein and butterfat into a delicious culinary delight that is both nourishing and long lasting — that is, if you can control yourself until your prize is properly aged. The cheeses in this chapter are characteristically firm or hard in texture, due to the removal of a high percentage of moisture during the cheese-making process. They are pressed and aged for varying lengths of time for full flavor development.

You will use two different cooking methods in these recipes. Just as commercial producers use water- or steam-jacketed vats to heat milk and curds, you will surround your pots with warm water. This method preserves the high acidity produced in the cheese-making process, resulting in strong-flavored cheeses. In other instances, as with washed-curd cheeses, whey is replaced with hot water to heat the curds from within the pot, resulting in less acidic, mild-flavored cheeses.

GOUDA, *page 206*

Following is an overview of the steps involved in hard-cheese making. For a thorough description of each of the techniques used in these recipes, see chapter 3. I recommend starting with the recipes here and keeping detailed notes so that with time and experience you can experiment with subtle variations and produce recipes for your own unique cheeses.

RIPENING

Ripening is the essential first step, whereby a bacterial starter triggers the conversion of milk sugar into lactic acid. The mixture is heated to speed the bacterial action. All but a few of these recipes call for using indirect heat (look for the icon at the top of the page). The increase of acid in the milk aids in the expulsion of whey from the curd, helps rennet coagulate the milk, helps preserve the final cheese, and aids in flavor development.

CALCIUM CHLORIDE

Calcium chloride restores the calcium balance in milk and aids in the renneting process. Due to seasonal variations in milk, differing fat content, and the effects of cold storage and pasteurization, I recommend adding calcium chloride to all hard and semi-hard cheese recipes.

COLORING

Although coloring is not necessary in cheese making, several of the cheeses in this chapter are traditionally deeper in color than a typical cow's-milk cheddar. You may add annatto cheese coloring to re-create those hues in your cheeses. Take care to thoroughly mix the coloring into the milk before adding rennet, or the color will be patchy or streaky.

RENNETING

Having ripened the milk and regulated it to the renneting temperature, the next step is to add diluted rennet. If rennet is not diluted, it will be unevenly distributed in the milk, which can produce a faulty curd. Rennet coagulates milk protein into a solid mass of curd. Trapped within this mass are the butterfat and whey. After stirring well, keep the pot covered to help maintain a uniform temperature so the milk can coagulate evenly.

CUTTING THE CURD

Cutting the curd into small, uniform pieces helps remove most of the whey that is trapped inside. It is important that the curd be cut when ready, or as soon as the curd gives a clean break, and not too soon (when it is too soft to work with) or too late (when it will be too firm and prone to being crushed).

COOKING THE CURD

When making hard cheese, you want to remove much of the moisture from the curds so that the cheese can be safely preserved. Increasing the temperature of the curds helps remove more whey. Heating also helps firm the curd and increases acidity.

DRAINING

Draining after cooking removes more moisture and any remaining whey or wash water (as in washed-curd cheeses, such as Gouda). Reserve the whey for making ricotta and other whey cheeses (see page 254).

MILLING AND SALTING

Carefully breaking the curd into smaller pieces facilitates the even distribution of salt throughout the cheese, for the most uniform flavor and preservation (salt hinders the growth of undesirable bacteria).

MOLDING AND PRESSING

Once the curds are packed into the desired mold, they're pressed to force out more whey and give the cheese its final texture and shape. Be sure to increase the pressure gradually over the specified time.

AIR-DRYING/BANDAGING

The cheese is air-dried to form a protective rind in anticipation of the aging process. Alternatively, it may be bandaged and larded to keep its shape, preserve its coat, and prevent excessive loss of moisture through evaporation (see page 49).

WAXING/OILING

Waxing is helpful for certain cheeses that do not have a thick rind, as it keeps the cheese from drying out and retards the growth of mold. While not as protective, rubbing the cheese with a light coating of vegetable oil can also prevent further moisture loss.

AGING

During this stage, the full flavor and body of the cheese are allowed to develop. This is very important and requires proper temperature and humidity. Most varieties of hard cheeses ripen best at 55 to 65°F (13 to 18°C). The longer a hard cheese ages, the stronger its flavor becomes.

Note: U.S. law requires raw-milk cheeses to be aged for more than 60 days if they will be sold. Therefore, I must advise you to use pasteurized milk for all cheeses that will be aged for fewer than 60 days.

WITH MARIA SCHUMANN AND HER HUSBAND, JOSH KARP, music and cheese go hand in hand. The two met when they were singers with her father's renowned Bread and Puppet Theater (established in 1963), where Maria also took over milking the cows when the resident puppeteers went on tour during the winter. "I had all this milk and needed something to do with it, so a friend lent me a copy of *Home Cheese Making* and I started experimenting with cheese. That's when I fell in love with it."

After marrying and having their first son, the couple moved about 10 miles away to Josh's family farm to raise sheep for making cheese. Maria says she was in for a shock. "Unlike those cows, our sheep hated being milked and dried off the second I took their lambs away. Eventually it got a little easier, and now I milk anywhere from 35 to 50 ewes — the most my land can handle."

Initially they made cheese for themselves in an old part of the shed that Josh, a skilled carpenter, could repair as needed. When they decided to turn the farm into a business, they funded a new facility through grants and a personal loan. "The only way we could justify spending that much money on a creamery is that we are also an organic apple orchard and we keep bees for making honey, so the dairy is a multiuse facility. It's not just my sheep that have to pay back the investment."

There's still the never-ending question of quantity, since sheep give so little milk to begin with — which explains why sheep's milk is still a rarity in the United States, and at least one reason Maria chose sheep over cows (less competition). "I also love the flavor and qualities of sheep's milk and the cheeses made from it." So why not make feta, the classic sheep's-milk cheese? Explains Maria, "My neighbors at Bonnieview already make an awesome feta."

Music still plays an important role at the farm. "We do a fair amount of ritual events, celebrating the summer solstice by singing songs to our sheep, and in the winter we sing traditional songs to the apple trees." They also hope to resume hosting their annual Sing and Farm retreats. Music and cheese, together in perfect harmony — a match made in Vermont.

A CHEESE MAKER'S STORY

Maria Schumann
CATE HILL ORCHARD
Greensboro, Vermont

LEICESTER,
page 154

BRITISH-STYLE CHEESES

Cheddar is one of the most popular cheeses in the world. Its name is taken from a small village in southern England, and its origins go as far back as the late 1500s. Legend has it that a Stone Age skeleton was found in the ancient caves overlooking Cheddar Gorge, and on the wall above its head hung a vessel containing goat's milk. Over thousands of years, the milk had hardened into a peculiar substance, pleasant to the taste — the first cheddar cheese.

In colonial America, cheddar was one of the most common cheeses made by farm wives. The original recipes are somewhat involved and time consuming; however, you will find the results to be well worth the effort. The cheeses included in this section all belong to the cheddar family and involve the same basic steps ("to cheddar" has become a technique, so the name is no longer capitalized). Unlike Traditional Cheddar (page 147), both Farmhouse Cheddar (page 142) and Stirred-Curd Cheddar (page 144) do not require the process of cheddaring (cutting the drained curd into strips and allowing them to set at 100°F [38°C] for 2 hours), thus saving a lot of time.

Indeed, Farmhouse Cheddar is a good place to start when making your first cheddar. Derby and Leicester are two English cheeses with a taste and texture very similar to those of cheddar. Cheshire is a clothbound cheese that's dense and crumbly and bears the distinctive orange hue of annatto.

STIRRED–CURD CHEDDAR

When you're "pressed" for time, this cheese is a perfect alternative to Traditional Cheddar (page 147). You can make it with either cow's or goat's milk. For some delicious flavor variations, see page 146.

2 gallons milk

½ teaspoon calcium chloride diluted in ¼ cup cool, nonchlorinated water

1 packet direct-set mesophilic starter culture

4 drops annatto cheese coloring diluted in ¼ cup water (optional)

½ teaspoon liquid rennet (or ½ rennet tablet) diluted in ¼ cup cool, nonchlorinated water

2 tablespoons cheese salt

Yield: 2 pounds

1. Heat the milk to 90°F (32°C). Add the calcium chloride solution and stir well to combine. Sprinkle the starter over the surface of the milk, wait 2 minutes for the powder to rehydrate, then stir well. Cover and allow the milk to ripen for 45 minutes.

2. Add the diluted annatto coloring, if desired, stirring to distribute evenly.

3. Make sure the milk's temperature is 90°F (32°C). Add the diluted rennet and stir gently with an up-and-down motion for 30 seconds. If using cream-line cow's milk, top-stir for several minutes longer. Cover and let set at 90°F for 45 minutes, or until the curd is firm and gives a clean break.

4. Cut the curd into ¼-inch cubes. Allow the curds to set for 15 minutes.

5. Heat the curds to 100°F (38°C), increasing the temperature by no more than 2 degrees every 5 minutes for a total of 30 minutes. Stir gently every 3 minutes to keep the curds from matting. Once the curds reach 100°F, maintain the temperature and continue stirring every 3 minutes for 30 minutes. Let set for 5 minutes.

6. Drain off the whey. Pour the curds into a large colander and drain for several minutes. Do not drain too long, or the curds will mat.

7. Pour the curds back into the pot and stir them briskly with your fingers, separating any curd particles that have matted.

8. Add the salt and mix in. Do not squeeze the curds.

9. Keep the curds at 100°F (38°C) for 1 hour, stirring every 5 minutes to avoid matting. The curds can be kept at temperature by resting the cheese pot in a sink or bowl full of 100°F water.

10. Line a 2-pound cheese mold with cheesecloth. Place the curds in the mold. Press the cheese at 15 pounds of pressure for 10 minutes.

11. Remove the cheese from the mold and gently peel away the cheesecloth. Turn over the cheese, re-dress it, and press at 30 pounds of pressure for 10 minutes.

12. Unwrap, flip, rewrap, and press at 40 pounds of pressure for 2 hours.

13. Unwrap, flip, rewrap, and press at 50 pounds of pressure for 24 hours.

14. Remove the cheese from the mold and peel away the cheesecloth. Air-dry at room temperature for 2–5 days, or until the cheese is dry to the touch, turning twice a day and rubbing off any mold growth.

15. Oil, wax, or bandage the cheese (see pages 49 and 50).

16. Age the cheese at 45–55°F (7–13°C) for 2–6 months, turning weekly and removing unwanted mold.

Flavor Variations

Sage Cheddar

Boil 1–3 tablespoons chopped fresh sage (or 1–3 teaspoons dried) in ½ cup water for 15 minutes, adding more water to cover the herbs as needed. Strain the flavored water into a small bowl and let cool; reserve the boiled sage. Follow the recipe on page 144, heating the flavored water with the milk according to step 1 and gently stirring the reserved sage into the curds in step 8.

Caraway Cheddar

Boil ½–2 tablespoons caraway seeds in ½ cup water for 15 minutes, adding more water to cover the seeds as needed. Strain the flavored water into a small bowl and let cool; reserve the boiled seeds. Follow the recipe on page 144, heating the flavored water with the milk according to step 1 and gently stirring the boiled seeds into the curds in step 8.

Jalapeño Cheddar

Boil ½–4 tablespoons chopped jalapeño, depending on desired degree of spicy heat, in ½ cup water for 15 minutes, adding more water to cover the chiles as needed. Strain the flavored water into a small bowl and let cool; reserve the boiled peppers. Follow the recipe on page 144, heating the flavored water with the milk in step 1 and gently stirring the boiled peppers into the curds in step 8.

TRADITIONAL CHEDDAR

*Making cheddar the time-honored way takes longer but you will be rewarded
with a singularly delicious result.*

2 gallons milk

½ teaspoon calcium
chloride diluted
in ¼ cup cool,
nonchlorinated water

1 packet direct-set
mesophilic starter
culture

½ teaspoon liquid
rennet (or ½ rennet
tablet) diluted
in ¼ cup cool,
nonchlorinated water

2 tablespoons cheese
salt

Yield: 2 pounds

1. Heat the milk to 86°F (30°C). Add the calcium chloride solution and stir well to combine. Sprinkle the starter over the surface of the milk, wait 2 minutes for the powder to rehydrate, then stir well. Cover and allow the milk to ripen for 45 minutes.

2. Make sure the milk's temperature is 86°F (30°C). Add the diluted rennet and stir gently with an up-and-down motion for 30 seconds. If using cream-line cow's milk, top-stir for another 30 seconds. Cover and let set undisturbed for 45 minutes.

3. Cut the curd into ¼-inch cubes. Allow the curds to set for 5 minutes.

4. Over the next 30 minutes, heat the curds to 100°F (38°C), increasing the temperature 2 degrees every 5 minutes. Stir gently every 3 minutes to keep the curds from matting.

5. Once the curds reach 100°F (38°C), maintain the temperature and continue stirring the curds every 3 minutes for another 30 minutes.

6. Allow the curds to rest for 20 minutes.

7. Pour the curds and whey into a colander set over a large bowl. Place the colander of curds back into the pot and let rest for 15 minutes.

Recipe continues on next page

8. Remove the colander from the pot and place the mass of curd on a cutting board. Cut the curd into 3-inch slices. Pour out any whey that has collected and put the pot into a sink full of 100°F (38°C) water. Place the slices in the pot and cover the pot. Maintain the curds at this temperature, turning them by hand every 15 minutes for 2 hours, until they are tough, with a texture similar to raw chicken.

9. Break the slices into walnut-size pieces with your fingers and put them back into the covered pot, still sitting in a sink full of 100°F water. Stir the curds with your fingers every 10 minutes for 30 minutes. Do not squeeze the curds; merely stir them to keep them from matting.

10. Remove the pot from the sink. Add the salt and stir gently.

11. Line a 2-pound cheese mold with cheesecloth. Place the curds in the mold. Press for 30 minutes at 10 pounds of pressure.

12. Remove the cheese from the mold, unwrap, flip, and rewrap. Press for 1 hour at 40 pounds of pressure.

13. Remove the cheese from the mold, unwrap, flip, and rewrap. Press for 3 hours at 50 pounds of pressure.

14. Remove the cheese from the mold. Peel away the cheesecloth. Air-dry the cheese at room temperature for 2–5 days, or until it is dry to the touch.

15. Oil, wax, or bandage the cheese (see pages 49 and 50).

16. Age the cheese at 50–55°F (10–13°C) for 3–12 months.

For Queen Victoria's wedding celebration in 1840, the farmers in Somerset, England, made a cheddar cheese weighing 1,100 pounds and measuring more than 9 feet wide.

TRADITIONAL
CHEDDAR

CHESHIRE

Once known as one of Britain's finest cheeses, Cheshire is almost lost to history. There are only a few farms still making this deep orange cheese, so the best way to find it is to make it in your own kitchen.

4 gallons milk

1 teaspoon calcium chloride diluted in ¼ cup cool, nonchlorinated water

1 packet C101 direct-set mesophilic starter culture or ½ teaspoon MA11 mesophilic starter culture (for raw milk, use ⅜ teaspoon)

2 teaspoons annatto cheese coloring, diluted in ¼ cup cool, nonchlorinated water

1 teaspoon liquid rennet (or 1 rennet tablet) diluted in ¼ cup cool, nonchlorinated water

3½ tablespoons cheese salt

Yield: 4 pounds

1. Slowly heat the milk to 86°F (30°C). Add the diluted calcium chloride and stir well to combine. Sprinkle the starter over the surface of the milk, wait 2 minutes for the powder to rehydrate, then stir well.

2. Scoop a bit of the warm milk into a small bowl. Add the annatto coloring and mix well. Pour the mixture back into the pot and stir slowly for 30 seconds with an up-and-down motion. (You will not be able to see the color change at this point because of the water content.) Cover the pot and let the milk ripen at 86°F (30°C) for 45 minutes. The curds can be kept at temperature by resting the cheese pot in a sink or bowl full of 86°F water.

3. Add the diluted rennet and stir slowly with an up-and-down motion for 30 seconds. Cover and let set undisturbed for 60 minutes, until the curd gives a clean break. Try to keep the contents warm, but it's okay if the temperature drops a few degrees during this time.

4. Cut the curd into ½- to ¾-inch cubes. Stir gently, then let the curds settle in the whey for 5 minutes. Slowly heat the curds to 90°F (32°C) over 60 minutes, stirring intermittently.

5. After 60 minutes, allow the curds to settle to the bottom of the pot and wait 30 more minutes. Place a colander lined with butter muslin in a tray.

6. Gently ladle the curds into the colander, wrap them in the cloth, and add a 10-pound weight. Let the curds drain for 15 minutes. Unwrap the curd and break it into 3- or 4-inch pieces. Rewrap the curds in the cloth and return them to the colander. Let the curds drain for 3 hours, turning them every 10 minutes for the first hour and every 30 minutes for the last 2 hours. Try to keep the curds at 90°F (32°C) as they drain.

7. Break up the curds into ½- to ¾-inch pieces. Add the salt to the curds, stirring gently to coat them evenly. Once the curds have absorbed the salt, transfer them to a 2- to 4-pound cheese mold lined with butter muslin. Let the curds drain at a temperature of 75–80°F (24–27°C) for 10 hours, turning the mold occasionally.

8. Remove the cheese from the mold, rewrap it in butter muslin, and place it in a cheese press. Press the cheese for 2 days, beginning with a weight of 15 pounds and increasing gradually to 150 pounds as follows. Before each increase of weight, unwrap the cheese, flip it, and rewrap it.

   - 15 pounds for 2 hours
   - 35 pounds for 4 hours
   - 60 pounds for 6 hours
   - 100 pounds for 12 hours
   - 150 pounds for 24 hours

9. Remove the cheese from the press and dry the surface with sanitized cheesecloth. Oil, wax, or bandage the cheese (see pages 49 and 50). Age the cheese at 55–60°F (13–16°C) for 5–10 weeks. The early-ripened cheese will be ready at 5–6 weeks, but the longer the aging time, the more improved the flavor.

In 1801, the townspeople of Cheshire, Massachusetts, created a mammoth cheese in honor of Thomas Jefferson's presidential victory. The curd was produced on individual farms from one day's milk (900+ gallons) and then pressed on the village green in a cider press. The 1,200-pound cheese was transported by sled, boat, and wagon to the White House.

DERBY CHEESE

*With origins in Derbyshire, England, this cheese is very similar to cheddar but
ages more quickly thanks to its higher moisture content.*

2 gallons milk

½ teaspoon calcium
chloride diluted
in ¼ cup cool,
nonchlorinated water

1 packet direct-set
mesophilic starter
culture

½ teaspoon liquid
rennet (or ½ rennet
tablet) diluted
in ¼ cup cool,
nonchlorinated water

2 tablespoons cheese
salt

Cheese wax

Yield: 2 pounds

1. Heat the milk to 84°F (29°C). Add the calcium chloride solution and stir
   well to combine. Sprinkle the starter over the surface of the milk, wait
   2 minutes for the powder to rehydrate, then stir well. Cover and allow
   the milk to ripen for 45 minutes.

2. Make sure the milk's temperature is 84°F (29°C). Add the diluted rennet
   and stir gently with an up-and-down motion for 30 seconds. If using
   cream-line cow's milk, top-stir for 1 minute longer. Cover and allow to set
   at 84°F for 50 minutes.

3. Cut the curds into ½-inch cubes.

4. Heat the curds to 94°F (34°C), increasing the temperature no more than
   2 degrees every 5 minutes. Stir gently to keep the curds from matting.
   When the temperature reaches 94°F, stir the curds for 10 minutes.

5. Place the mass of curd on a draining board and cut it into 2-inch slices.
   Lay the slices on the draining board and cover them with a clean towel,
   which should occasionally be placed in a bowl of 94°F (34°C) water, then
   wrung out. (Never use a towel that has been used during bread making.
   The yeast can cross-contaminate the cheese, making it spongy.) The aim
   is to keep the temperature of the curd slices at 94°F. Turn over the slices
   every 15 minutes. Let them drain for 1 hour.

6. Break the curd slices into quarter-size pieces. The curd should be tough
   and tear when pulled apart. Add the salt and stir gently.

7. Place the curds in a cheesecloth-lined 2- to 4-pound cheese mold. Press
   for 10 minutes at 15 pounds of pressure.

8. Remove the cheese from the mold and gently peel away the cheesecloth. Turn over the cheese, re-dress it, and press for 2 hours at 30 pounds of pressure.

9. Unwrap, flip, rewrap, and press for 24 hours at 50 pounds of pressure, flipping and unwrapping halfway through.

10. Remove the cheese from the mold, peel away the cheesecloth, and air-dry at room temperature for 2–5 days, or until the cheese is dry to the touch.

11. Wax the cheese (see page 50).

12. Age the cheese at 50–55°F (10–13°C) and 80–85% relative humidity for 6 months.

    NOTE: Can be eaten at 4–6 weeks if you prefer a milder cheese.

VARIATION: SAGE DERBY

Sage Derby is one of England's oldest and most revered cheeses, dating back to the sixteenth century. It is increasingly hard to find. To make it the traditional way, fill the mold halfway with curds, sprinkle evenly with chopped fresh sage, then add the remaining curds. Or mix the sage into the curds for a mottled result.

LEICESTER

A mild, hard cheese similar to cheddar, Leicester (named for the English town where it was created) ripens somewhat more quickly. Although it typically has coloring added, you can omit it if you prefer.

2 gallons milk

½ teaspoon calcium chloride diluted in ¼ cup cool, nonchlorinated water

1 packet direct-set mesophilic starter culture

1 teaspoon annatto cheese coloring diluted in ¼ cup water (optional)

½ teaspoon liquid rennet (or ½ rennet tablet) diluted in ¼ cup cool, nonchlorinated water

2 tablespoons cheese salt, plus a pinch for dusting

Yield: 2 pounds

1. Heat the milk to 85°F (29°C). Add the calcium chloride solution and stir well to combine. Sprinkle the starter over the surface of the milk, wait 2 minutes for the powder to rehydrate, then stir well.

2. Add the diluted annatto coloring, if desired, stirring to distribute evenly. (You will not be able to see the color change at this point because of the water content.) Cover and allow the milk to ripen at 85°F (29°C) for 45 minutes.

3. Make sure the milk's temperature is at 85°F (29°C). Add the diluted rennet and stir gently with an up-and-down motion for 30 seconds. If using cream-line cow's milk, top-stir for several minutes longer. Cover and allow to set at 85°F for 45 minutes, or until the curds give a clean break.

4. Cut the curd into ¼-inch cubes. Stir gently every 3 minutes for 15 minutes to keep the curds from matting.

5. Heat the curds to 95°F (35°C), increasing the temperature no more than 2 degrees every 5 minutes. Maintain the curds at 95°F for 30 minutes, again stirring gently every 3 minutes.

6. Gently ladle the curds into a colander and let them drain for 20 minutes.

7. Place the mass of curd on a draining board. Cut it into 2-inch-long slices to drain. Lay the slices on the draining board and cover them with a clean towel, which should occasionally be placed in a bowl of 96°F (36°C) water, then wrung dry. (Never use a towel that has been used during bread making. The yeast can cross-contaminate the cheese, making it spongy.) Turn the slices every 20 minutes over 1 hour.

8. Break the slices into nickel-size pieces and put them in a bowl. Stir them for several minutes. Add the 2 tablespoons salt and gently stir for several minutes longer.

9. Place the curds in a cheesecloth-lined 2- to 4-pound cheese mold and press for 30 minutes at 15 pounds of pressure.

10. Remove the cheese from the mold and gently peel away the cheesecloth. Turn over the cheese, re-dress it, and press for 2 hours at 30 pounds of pressure.

11. Unwrap, flip, rewrap, and press for 24 hours at 50 pounds of pressure.

12. Remove the cheese from the mold and peel away the cheesecloth. Dust the cheese with salt and shake off any excess. Air-dry the cheese at room temperature for 2–5 days, or until it is dry to the touch.

13. Wax the cheese (see page 50).

14. Age the cheese at 55°F (13°C) for at least 12–16 weeks. For a very subtle, distinctive flavor, allow it to mature for 9 months.

FRENCH-STYLE CHEESES

There are several categorizations to help us bring order to the vast world of French cheese. Most French cheeses can fit into one of three categories: pressed cheeses, like Cantal and Comte; soft cheeses, like Brie and Camembert; and blue cheeses, like Roquefort and Bleu d'Auvergne. Of these, over 40 are protected by the AOC, a designation that ensures that only cheese made with the traditional recipe and method can claim the name.

Many French cheeses are made with unpasteurized milk, and this makes them impossible to import into the United States. More industrialized, pasteurized versions may be available in your local grocery or cheese shop, but those imported versions are far from the original. To taste many of these phenomenal cheeses, we can travel to the country and visit the farms that create them. Or, of course, we can make them ourselves at home.

CANTAL, *page 160*

TOMME, *page 158*

RACLETTE, *page 162*

Tomme–Style Cheese

Tomme is one of the great Alpine cheeses. It's a beautiful wheel with a rustic rind created by wild molds, a true representation of the mountains that create it. This is often made as a low-butterfat cheese, when the fat has been skimmed for butter or cream.

4 gallons milk

1 teaspoon calcium chloride diluted in ¼ cup cool, nonchlorinated water

1 packet C101 direct-set mesophilic starter culture

1 packet C201 direct-set thermophilic starter culture

½ teaspoon liquid rennet (or ½ rennet tablet) diluted in ¼ cup cool, nonchlorinated water

1 gallon Saturated Brine (page 46)

Cheese salt

Yield: 4 pounds

1. Slowly heat the milk to 87°F (31°C). Add the calcium chloride solution and stir well to combine. Sprinkle both starters over the surface of the milk, wait 2 minutes for the powders to rehydrate, then stir well. Cover the pot and let the milk ripen at 87°F for 1 hour.

2. Add the diluted rennet and stir with an up-and-down motion for 30 seconds. Cover the pot and let set undisturbed at 87°F (31°C) for 30 minutes, or until the curd gives a clean break.

3. Cut the curds vertically into 1-inch-square columns. Let the curd sit for 2–3 minutes, then use a ladle or wire whisk to cut the columns into ¼-inch pieces. Let the curd settle under the whey.

4. Remove about 1½ gallons of whey from the pot. Very slowly add just over ½ gallon of 120°F (49°C) water to the pot, slowly bringing the curd up to 98°F (37°C) over the course of 15 minutes. The final result will be a firm curd, fully cooked through and resistant to the touch. Let the curd settle under the whey.

5. Using a colander in the pot to hold down the curds, scoop out the whey until the liquid level is just at the level of the curds. Give the curd a good stir to prevent sticking. Use a ladle to transfer the curd and any remaining whey to two small cheese molds or a large Tomme mold lined with butter muslin.

6. Add 4–5 pounds of weight to the cheese and press, turning the cheese every 30 minutes until the cheese no longer weeps whey, 3–4 hours. Let the cheese rest without the weight at room temperature for 8–10 hours.

7. Remove the cheese from the mold and submerge it in the saturated brine for 8 hours. Sprinkle the exposed surface with salt. Flip the cheese halfway through the soak time and sprinkle the other side with salt. Remove the cheese from the brine and wipe the surface with a dry cloth. Let it dry at room temperature for 1–2 days, turning twice a day. The surface will darken over this time.

8. Age the cheese at 52–56°F (11–13°C) and 80–85% relative humidity for 6–8 weeks, gently rubbing the mold into the surface and turning the cheese every few days. The character of this cheese comes from the wild molds on the rind, so you will see many different kinds of flora growing on your cheese over this time. Typically the coating will start as a furry mucor mold along with a white mold, which will darken as the cheese is flipped and rubbed. The mold may become mottled with some spots of mimosa yellow or bright red — these are quite typical and a sign of good, diverse flora in your aging space. Continue to gently rub the surface, but let the molds do their work.

CANTAL

Cantal is one of the oldest cheeses of France, and its sweet and rich flavor is reminiscent of English-style cheddar. In Auvergne, it's only made with winter milk from the Salers breed of cow. The summer milk of the same cows is used to make Salers cheese.

7 gallons milk

1¾ teaspoons calcium chloride diluted in ¼ cup cool, nonchlorinated water

⅜ teaspoon MA4002 mesophilic and thermophilic blend starter culture or 1 packet C101 direct-set mesophilic starter culture

⅔ teaspoon liquid rennet (or ⅔ rennet tablet) diluted in ¼ cup cool, nonchlorinated water

Cheese salt

Yield: 7 pounds

1. Slowly heat the milk to 90°F (32°C). Add the calcium chloride solution and stir well to combine. Sprinkle the starter over the surface of the milk, wait 2 minutes for the powder to rehydrate, then stir well. Cover the pot and let the milk ripen at 90°F for 30 minutes.

2. Add the diluted rennet and stir gently with an up-and-down motion for 30 seconds. Cover the pot and allow the curd to set at 90°F (32°C) for 60 minutes, or until the curd gives a clean break.

3. Cut the curd into ½-inch cubes. Stir slowly, maintaining a temperature of 90°F (32°C), until the curds shrink and firm up, 20–30 minutes. Let the curds settle to the bottom of the pot for 5 minutes. Using a colander in the pot to hold down the curds, scoop out the whey until the liquid level is 1 inch above the curds. Place a plate on the curd and top it with a 25-pound weight for 5 minutes. Remove the rest of the whey as it is expelled. Wrap the curd in butter muslin and transfer it to a colander. Reapply the weight, and let it sit and drain for 30 minutes.

4. Unwrap the curd and cut it into 2-inch strips. Turn the strips over, rewrap the curd, and increase the weight to 35 pounds. Repeat this process 3–6 times, increasing the weight each time by 10 pounds, until the curd is quite dry. Try to keep the curd at 80–90°F (27–32°C) through the draining process. Break up the curd mass into rough blocks and let them ripen at room temperature for 8–10 hours.

5. Break up the rough blocks into 1-inch pieces. Sprinkle 2½ teaspoons of salt over the curds, stirring gently to distribute it evenly.

6. Line a large hard-cheese mold with butter muslin and fill it with the salted curds. Top the mold with a follower, and press the cheese at the following weights. With each increase of weight, unwrap the cheese, flip it, and rewrap it.

   - 25 pounds for 3 hours
   - 50 pounds for 6 hours
   - 90 pounds for 12 hours
   - 150 pounds for 20 hours
   - 250 pounds for 24 hours

7. Remove the cheese from the mold, and age for 3–9 months at 54°F (12°C) and 80–85% relative humidity. Dry-brush the rind regularly as it develops.

How can you govern
a country which has
246 varieties of cheese?

CHARLES DE GAULLE

RACLETTE

Raclette is an Alpine cheese with a long history. The cheese got its name from the word racler, *which means "to scrape." Cheese makers in the mountains would hold the cut face of the wheel to the fire until it melted enough to be scraped onto bread or boiled potatoes. Raclette is still eaten this way in Alpine establishments. This cheese is best made with freshly milked raw milk.*

7 gallons raw milk, skimmed to 3–3.5%

1¾ teaspoons calcium chloride diluted in ¼ cup cool, nonchlorinated water

1 packet C21 buttermilk starter culture

4½ ounces fresh yogurt prepared from Y1 yogurt culture

Just under ½ teaspoon liquid rennet (or ½ rennet tablet) diluted in ¼ cup cool, nonchlorinated water

1 gallon Saturated Brine (page 46)

Cheese salt

Light Brine (page 46)

Brevibacterium linens (red bacteria)

Yield: 7 pounds

1. Slowly heat the milk to 90°F (32°C). Add the calcium chloride solution and stir well to combine. Sprinkle the buttermilk starter over the surface of the milk, wait 2 minutes for the powder to rehydrate, then stir well. Stir in the Y1 yogurt. Cover the pot and let the milk ripen for 30 minutes at 90°F.

2. Add the diluted rennet and stir gently with an up-and-down motion for 30 seconds. Cover and let set undisturbed at 90°F (32°C), or until the curd is just firm enough to cut, about 40 minutes.

3. Cut the curds vertically into 1-inch columns. Let the curd rest for 5 minutes as the edges of the cuts firm up. Then make the second cut, creating ¼-inch cubes in the curd. You can use the knife for the second cut, or switch over to a wide wire whisk.

4. For a drier cheese, slowly heat up the curds to 100°F (38°C) over the next 25–30 minutes, stirring gently every 3 minutes to prevent the curds from matting. Hold the curd at 100°F, stirring for an additional 15 minutes. The final result should be firm curd, fully cooked through and moderately resistant to the touch. Alternatively, for a moister and sweeter cheese, you can remove one-quarter of the whey. Slowly add enough 140°F (60°C) water over the course of 30 minutes to bring the curd up to 100°F. Then stir for 15 minutes. Let the curds settle under the whey.

5. Remove enough whey from the pot so that the liquid level is just above the curd level. Use a mesh ripening mat to help to consolidate the curd, wrapping it around the curd and drawing it together. Roll the full

curd mass in a length of butter muslin, grab the cloth by the corners, and quickly drop it into a large hard-cheese mold. Place a follower on top of the curd, and weight it with 25 pounds. Let the cheese rest in a warm space (80–86°F/27–30°C) for 1 hour. Then unwrap the cheese, flip, rewrap, and return to the mold with 25 pounds of weight for another hour.

6. Unwrap the cheese, flip, rewrap, and return it to the mold. Press for 12 hours with 50 pounds of weight. During this time, the cheese will cool to room temperature.

7. Remove the cheese from the mold and submerge it in the saturated brine for 14–16 hours. Sprinkle the exposed surface with salt. Flip the cheese halfway through the soak time and sprinkle the other side with salt.

8. Wipe the surface of the cheese and let it dry out for 1–2 days, turning twice daily. The surface will darken over this time.

9. Transfer the cheese to an aging space at 52–54°F (11–12°C) and 90–94% relative humidity. The humidity will be easier to maintain if you store the cheese in a covered container or cake carrier. If any mold develops over the first few days, simply wipe it away with a dry cloth.

10. After 2 to 3 days, the surface will become greasy and a thin slime layer will appear. You might also notice a fruity aroma. Turn the cheese over, and wipe the top and sides to remove any mold. Brush the surface of the cheese with light brine every day for the first 10 days, adding a pinch of *B. linens* to the brine the first two times. Let it dry with the dry side down. Once the surface mold starts to develop, wash it twice a week with light brine until the cheese is ripe, 3–4 months. Eventually the cheese will have a rosy appearance with a thin rind and a pungent aroma.

NOTE: If the wash is too frequent, a thick crust may build up; if allowed to dry, it may exfoliate (crack and peel).

REBLOCHON–STYLE CHEESE

Since as far back as the thirteenth century and for many centuries after, Reblochon was the illegal cheese of the farms of the Haute-Savoie region of France. It was the secret product of animals milked outside the jurisdiction of the government, a rich cheese shared only with family and friends. When the French Revolution brought freedom to the tenant farmers, the cheese lost its contraband status and instantly became one of the most popular cheeses in France.

Due to the FDA's rules on pasteurization, it is now illegal to sell Reblochon, which is made with raw milk, in the United States, so the only way to enjoy this wonderful cheese is to make it yourself. You'll need a dedicated pine or fir aging board for this cheese, washed only in water and left in the sun to bleach for a day before use.

4 gallons raw milk

1 teaspoon calcium chloride diluted in ¼ cup cool, nonchlorinated water

⅛ teaspoon Thermo B starter culture or 1 ounce prepared Bulgarian yogurt from a Y1 yogurt culture

⅛ teaspoon MM100 mesophilic starter culture or 1 packet C21 buttermilk starter culture

¹⁄₁₆ teaspoon *Geotrichum candidum*

¹⁄₃₂ teaspoon *Brevibacterium linens*

¼ teaspoon liquid rennet (or ¼ rennet tablet) diluted in ¼ cup of cool, nonchlorinated water

1 gallon Saturated Brine (page 46)

Cheese salt

Light Brine (page 46)

AGING MIXTURE (FOR STEP 9)

1 cup sterilized water

1 tablespoon cheese salt

Pinch of *G. Candidum*

Pinch of *B. linens*

Pinch of sugar

Note: If using pasteurized milk, double the amount of the two cultures and the rennet.

Yield: 4 pounds

1. Slowly heat the milk to 94°F (34°C). Add the diluted calcium chloride and stir well to combine. Sprinkle the Thermo B starter, mesophilic starter, *G. candidum*, and *B. linens* over the surface of the milk, wait 2 minutes for the powders to rehydrate, and stir well. (If you are using yogurt instead of Thermo B, add it after the powders rehydrate and stir well.) Cover the pot and let the milk ripen for 60 minutes at 94°F.

2. Add the diluted rennet and stir with an up-and-down motion for 30 seconds. Cover and let set undisturbed at 94°F (34°C) for 20 minutes, or until the curd gives a clean break.

3. Use a long knife to cut the curd into ½-inch cubes. Take your time, aiming to complete the task in 15–20 minutes. Then stir the curd very gently for 5–10 minutes. Collect a thin layer of curds in your palm, let it drain for a few seconds, then flip your hand over. If the curds stick, move on to the next step. If they fall back into the pot, stir for another 5 minutes and test the curd again. Let the curds settle to the bottom of the pot.

4. Set five 5- to 6-inch Tomme molds on draining mats. Arrange the molds in a cluster so you can use a single piece of butter muslin to line all of them if you choose to.

5. Remove enough whey from the pot so there is only 1 inch of whey above the curds. Using a ladle, quickly transfer the curd into the molds, mounding it above the top of each mold. As soon as all the curd is in the molds, flip each cheese. The cheeses will already have a good consolidation and a fairly smooth outer surface. Add 3 pounds of weight (a quart jar filled with water works well) and a follower to each cheese. Let them drain, turning twice over the next 30 minutes.

6. Unwrap the cheeses from the butter muslin and return them to the molds without the cloth. Add the 3-pound weights and the followers and let the cheeses drain at 75°F (24°C) for 3–4 hours. Remove the weights and leave the cheeses in the molds for 5 hours. (If you are checking pH, they will have a final pH of 5.3–5.4.)

Recipe continues on next page

7. Remove the cheeses from the molds and place them in the saturated brine for 1½ hours. Sprinkle the exposed surface with 1–2 teaspoons of salt. Flip the cheese and sprinkle the other side with salt halfway through the brine time. Remove the cheeses from the brine bath and let them air-dry at room temperature for about 8 hours, turning once.

8. Transfer the cheeses to their dedicated aging board in an aging space at 58–62°F (14–17°C) with 92–95% relative humidity. The humidity will be easier to maintain if you age the cheese in a covered container or cake carrier. Let the cheese age until it emits a fruity aroma and the surface has a greasy feel, 2–3 days.

9. While the cheese rests, on day 2 make the aging mixture and let it sit overnight at 72°F (22°C) to develop.

10. On day 3 or 4 (when you smell the fruity aroma and the surface feels greasy), use a sanitized cloth to wipe the surface of the cheese with light brine. Then wipe the top and sides with the aging mixture. The next day, flip the cheese and repeat with the mixture. Watch carefully; don't allow the surface to get too slimy, too dry, or too wet. Continue to age the cheese, flipping it daily without applying the mixture. On day 7 you should see the white *Geotrichum* mold develop, and on day 9, the mold growth should be well established.

11. Move the cheese to an area with a temperature of 52–56°F (11–13°C) and 90% humidity. Wipe the top and sides of the cheese with the aging mixture. The next day, flip the cheese and wipe it again. Let the cheese rest, turning it daily, for 4 more days.

12. On day 14, when the molds are well established, wrap the cheeses in a washed-rind wrap, keeping them in the aging room. If the cheeses are brought to room temperature, the wrap has a tendency to stick. Continue to age at 52–56°F (11–13°C), turning them daily, for 30–45 days, depending on the degree of desired ripeness. The humidity does not have to remain so high because the wrap will keep the cheese moist inside.

IT'S NO WONDER THE CHEESES FROM SEQUATCHIE COVE CREAMERY
garner so many accolades: Their 45 cows roam 150 acres, grazing on the
plentiful grass, resting in the shade of the woods, and drinking from a river
that runs along the border of the pasture. The cheeses are named to evoke
the landscape: for example, Cumberland for the plateau where the farm
is located and Coppinger for the cove that sits below. "We want to make
something that doesn't exist anywhere else," explains Padgett, co-owner of
the creamery with her husband, Nathan. "There's a running elemental component of Sequatchie Cove flavors in all our cheeses."

Unlike proprietors of many other farmstead creameries, Padgett and
Nathan lease rather than own their land, which they had been using to
grow vegetables for a CSA since 2003. But the farm proved too rocky to
plow, so they had to figure out another way to use the available resources.
Since the owner was a dairy farmer and the land was ideal for growing
grass, their thoughts turned to cheese.

By 2006, Nathan had bought a couple of cows and was experimenting
at home before spending the next four years attending workshops and
studying with cheese makers. Soon after the creamery was up and running,
in 2012, they took home their first award for Dancing Fern, a Reblochon-
style cheese. "We had only made four or five batches, yet the expectation
was set. That cheese is still ridiculously hard to make. But I'm so happy that
happened, it changed everything."

Not that they haven't experienced some severe setbacks. In 2016, after
a record-breaking year of cheese production, they were hit with the worst
drought in decades, leaving the farm without enough grass to feed the cows,
and them without enough milk to make cheese. At the same time, the couple took over management of the herd when the owner decided to retire.

What's kept the creamery going, Padgett says, are people in their community who believe in their mission of a sustainable, farm-based, value-
added product. "We want to be able to raise our family on this beautiful
land. We don't own this farm and never will, but that's why we were even
able to do something like this, because someone else's farm was available."

A CHEESE MAKER'S STORY

Nathan and Padgett Arnold
SEQUATCHIE COVE CREAMERY
Sequatchie, Tennessee

CACIOCAVALLO,
page 176

ITALIAN-STYLE CHEESES

RICOTTA SALATA,
page 179

PARMESAN,
page 180

What helps define cheeses with Italian origins is the way they pair so adeptly with a great many foods, especially, of course, those from their home country, but also crossing over into "all-purpose" territory. Most use a thermophilic (heat-loving) starter in their preparation and, with the exception of the pulled-curd varieties, provolone and caciocavallo, follow the same basic steps as other hard cheeses.

Parmesan is, by far, the most familiar and beloved of Italian cheeses, but there are others worth exploring, too. How long you choose to age them depends on your preferences — and your patience. Asiago and Tuscan Pepato are ready to eat in as little as a month, Montasio in two months. Others (Romano, Parmesan) are usually aged for a minimum of five to 10 months, when the texture will be ripe for slicing. However, if you can muster the willpower to save one wheel for a full two years, both you and your fortunate friends will be in for quite a cheese revelation.

ASIAGO

Traditional aged Asiago has a sharp flavor, reminiscent of Parmesan or Grana Padano. This younger version, also known as Asiago Pressato, is far milder. The short aging period creates a cheese with a supple texture and bright flavor, with an aroma of yogurt and rich butter. This recipe uses two different thermophilic cultures. The first converts the lactose in the milk to lactic acid, and the second transforms only part of the milk sugar, leaving a sweet note in the final cheese. This recipe can be scaled up or down, as long as the ingredients are adjusted proportionally.

6 gallons milk

4 ounces heavy cream, for a richer cheese (optional)

1½ teaspoons calcium chloride diluted in ¼ cup cool, nonchlorinated water

1 packet C201 direct-set thermophilic starter culture or ⅛ teaspoon TA61 thermophilic starter culture

1⁄16 teaspoon LH100 thermophilic starter culture

1 teaspoon liquid rennet (or 1 rennet tablet) diluted in ¼ cup cool, nonchlorinated water

1 gallon Saturated Brine (page 46)

Cheese salt

Light Brine (page 46)

Yield: 6 pounds

Note: If you're making this cheese in the winter with a more fatty winter milk, increase your milk temperatures (both for culturing and cooking the curd) 5 degrees over the those given in this recipe.

1. Slowly heat the milk and heavy cream, if desired, to 95°F (35°C). Add the calcium chloride solution and stir well to combine. Sprinkle both starters over the surface of the milk, wait 2 minutes for the cultures to rehydrate, then stir well. Cover the pot and let the milk ripen for 30 minutes at 95°F.

2. Add the diluted rennet and stir with an up-and-down motion for 30 seconds. Cover and let set undisturbed at 95°F (35°C) for 25 minutes, or until the curd gives a clean break.

3. Cut the curd into ⅜- to ½-inch cubes. Gently stir the curds as you heat them back up to 95°F (35°C) if the temperature dropped. Continue to stir until the curds become firm, about 15 minutes.

4. Gradually heat the curds to 106°F (41°C) over 20 minutes, stirring gently to keep the curds from matting. Continue to stir slowly for another 15 minutes.

5. Heat the curds to 118°F (48°C) over 10 minutes while stirring slowly. After 10 minutes, allow the curds to settle to the bottom for 20 minutes, stirring gently every 3 minutes. The curds should now be dry enough that they retain their shape when pressed between your fingers with moderate pressure.

6. Using a colander in the pot to hold down the curds, remove the whey to the level of the curds.

7. Use a skimmer to transfer the curds to a large hard-cheese mold lined with cheesecloth.

8. Press the cheese at 12 pounds of pressure for 30 minutes. Unwrap the cheese, flip, rewrap, and return it to the mold. Press with 25 pounds of weight for 2 hours. Unwrap, flip, rewrap, and press with 25 pounds of weight for another 2 hours. Unwrap, flip, rewrap, and return to the mold. Transfer the mold to a warm place (75°–85°F/24–29°C) and let it sit undisturbed for 36 hours.

9. Remove the cheese from the mold and let it cool to room temperature. Place the cheese in the saturated brine and sprinkle the exposed surface with salt. Let rest for 9 hours. Flip the cheese and sprinkle the other side with salt. Let rest for another 9 hours. Remove the cheese from the brine and pat dry with sanitized cheesecloth.

10. Age the cheese at 54–58°F (12–14°C) and 85% relative humidity for 30–40 days. Wipe off any surface mold with light brine.

GRANA PADANO

This is a Parma-type cheese, quite similar to Parmigiano-Reggiano. The difference is in the requirements of the milk, which are more relaxed than those for Parmigiano-Reggiano. Grana Padano is often aged less than Parmigiano-Reggiano, although it's possible to find both young and aged versions.

Toscano Pepato

This is a delicious spicy cheese with a kick, based on Vacha Toscano, a cow's-milk cheese made in the south of Tuscany. It gets its spice from smoked jalapeños and cracked peppercorns, both added as you build the curd in the molds.

5 gallons milk

1½ teaspoons calcium chloride diluted in ¼ cup cool, nonchlorinated water

⅛ teaspoon MA11 mesophilic starter culture

¼ teaspoon TA61 thermophilic starter culture

¾ teaspoons liquid rennet (or ¾ rennet tablet) diluted in ¼ cup cool, nonchlorinated water

2 dried smoked jalapeños, rehydrated, seeded, and roughly chopped

⅛ cup peppercorns, roughly chopped

1 gallon Saturated Brine (page 46)

Cheese salt

Light Brine (page 46)

Yield: One 5-pound round

1. Slowly heat the milk to 98°F (37°C). Add the calcium chloride solution and stir well to combine. Sprinkle the starters over the surface of the milk, wait 2 minutes for the powders to rehydrate, then stir well. Cover the pot and let the milk ripen for 1 hour at 98°F.

2. Add the diluted rennet and stir with an up-and-down motion for 30 seconds. Cover and let set undisturbed at 98°F (37°C) for 20 minutes, or until the curd gives a clean break.

3. Cut the curd vertically into ¾-inch columns and let it sit for 5 minutes. Use a wire whisk to break the curd into coffee bean–sized pieces. Heat the curds to 118°F (48°C) over 30 minutes, stirring gently every 3 minutes. Continue stirring the curds for about 20 more minutes. The final curds will retain their shape when pressed between your fingers with moderate pressure.

4. Remove the whey from the pot until the liquid level is just to the curd level. Use a ladle to transfer the curd into a large Tomme mold lined with butter muslin, building a ½-inch layer of curd in each form. Then add a thin layer of chopped jalapeños and peppercorns, leaving a ½-inch band

of plain curd around the edge. (This will help the finished cheese encapsulate the peppers.) Continue to alternate layers of curd and pepper mixture, completing the process with a layer of curd.

5. Keep the curd at 90–95°F (32 35°C) as you press it. Top the mold with a follower, and press the cheese at the following weights. With each increase of weight, unwrap the cheese, flip it, and rewrap it.

   • 8 pounds for 60 minutes
   • 25 pounds for 4–6 hours

6. Remove the weight from the cheese, and let it rest at room temperature for 8–10 hours. Remove the cheese from the mold and place it in the saturated brine for 8 hours. The cheese will float, so sprinkle the top surface with salt. Flip the cheese halfway through the soak time, and sprinkle the other side with salt. Remove the cheese from the brine and dry off the surface with a cloth. Let the cheese rest at 52–58°F (11–14°C) and 90–95% relative humidity until it develops a rind, about 1 week, flipping daily.

7. Age at 52–58°F (11–14°C) and 85% relative humidity for 1 month for a moist cheese and up to a year for a drier, more complex cheese. If any mold develops on the rind during aging, wipe it away with light brine.

Wine and cheese are ageless companions, like aspirin and aches, or June and moon, or good people and noble ventures.

M. F. K. FISHER

Provolone

Provolone is a pasta filata (pulled) type of cheese. According to tradition dating back to Roman times, provolone is ripened and smoked. Today, a milder variety with a delicate, light taste is widely produced. Provolone, from the Italian for "large oval" or "large sphere," may be molded into a variety of shapes and tied with a cord to leave an impression in the rind.

3 gallons milk (not ultrapasteurized)

½ packet direct-set C101 mesophilic starter culture

6 ounces prepared Bulgarian yogurt

¼ teaspoon lipase powder dissolved in ¼ cup cool water and allowed to sit for 20 minutes (optional)

⅜ teaspoon liquid rennet diluted in ¼ cup cool, nonchlorinated water

1 gallon Saturated Brine (page 46)

Light Brine (page 46)

Yield: About 3 pounds

1. Heat the milk to 90°F (32°C). Sprinkle the starter over the surface of the milk, wait 2 minutes for the powder to rehydrate, then stir well. Add the Bulgarian yogurt and stir well to combine. Cover and allow the milk to ripen for 30 minutes at 90°F.

2. Add the lipase solution, if using, and stir well. Let set for 10 minutes.

3. Add the rennet solution and stir gently with an up-and-down motion for 30 seconds. Let set for 30 minutes, or until you have a clean break.

4. Cut the curds into ½-inch cubes. Stir gently every 3 minutes for 10 minutes.

5. Over the next 45 minutes, heat the curds 3–5 degrees every 5 minutes to 112–118°F (44–48°C), stirring gently every 3 minutes to prevent the curds from matting. Let set for 5 more minutes, then remove the whey to the level of the curd.

6. Ladle the curds into a colander set into a pot of 105°F (41°C) water and let sit for 1 hour, covering the colander to keep the curds warm. Cut the curds into 1-inch slabs and allow to consolidate in the colander another 2–3 hours. If checking, the final stretching pH is 5.2–5.37.

7. Start your stretch test by taking a small piece of curd and dipping it into 180°F (82°C) water. If it does not stretch, try again at 15-minute intervals. Once this small piece has a smooth and shiny stretch, slice the rest of the curd into ¼-inch slices and submerge them into 180°F water. Let sit in the water until the entire mass stretches easily.

8. Working with gloved hands (or dipping your hands in cold water before handling), stretch the curd, forming it into one large ball as you go. Holding the ball in your hands, form it into a log by making a depression in the top surface with your thumb while pushing the excess curd into the center, shape the curd by pulling up on the mass from the bottom to the top and working it back down into the center hole. (If the curd cools down too much, it will not stretch; dip it back into the hot water.) As the curd stretches and becomes shiny, seal the end in the hot water and roll into a smooth log.

9. Once your cheese is completed, place it into a bowl filled with 50°F (10°C) water and let it sit for 5 minutes, then transfer to a bowl filled with ice water and let sit for 15–30 minutes, until cold. Soak in the saturated brine for 12 hours.

10. Remove the cheese from the brine and pat dry. Tie with a cord for hanging and air-dry at room temperature for 1–2 days.

11. Hang and age the cheese for 2 weeks at 52–58°F (11–14°C) in 80–85% relative humidity. At 2 weeks rub the cheese with oil and continue hanging for 4–9 months, depending on the desired sharpness. During aging, wipe any mold off with the light brine. (The cheese may also be cold-smoked at 40°F/4°C after brining and before hanging to age.)

CACIOCAVALLO

Caciocavallo is a traditional southern Italian stretched-curd cheese made from cow's, sheep's, or water buffalo's milk. It's known for its shape — similar to a gourd or a teardrop with a knot at the top. After three months of aging, caciocavallo is a creamy cheese similar to provolone, but after two years, it transforms into a sharper grating cheese. This recipe is easy to double as long as the amounts of the ingredients are proportional.

2 gallons milk (*not* ultrapasteurized)

1 packet C101 direct-set mesophilic starter culture or ⅟₁₆ teaspoon MA11 mesophilic starter culture

½ teaspoon liquid rennet diluted in ¼ cup cool, nonchlorinated water

1 gallon Saturated Brine (page 46)

Note: For a spicy cheese, add chopped smoked jalapeños or other peppers to the curd between removing the whey and molding.

Yield: 2 pounds

1. Slowly heat the milk to 92°F (33°C). Sprinkle the starter over the surface of the milk, wait 2 minutes for the powder to rehydrate, then stir well. Cover the pot and let the milk ripen for 30 minutes at 92°F.

2. Add the diluted rennet and mix well with an up-and-down motion for 30 seconds. Cover and let set undisturbed for 60 minutes. It's okay if the temperature drops a few degrees during this time.

3. Cut the curd into ½- to ⅝-inch cubes. The size of the curd will determine the moisture level of the finished cheese (smaller = drier, larger = moister).

4. Over the next 20 minutes, stir gently every 3 minutes while slowly heating the curds up to 102°F (39°C) by increasing the temperature 3–5 degrees every 5 minutes. Hold the curds at temperature and continue to stir until they're firm throughout and resist when pressed with your fingers. When done, remove the pot from the heat and let the curds settle for 5 minutes.

5. Using a colander in the pot to hold down the curds, scoop out the whey until the liquid level is just at the level of the curds. (This whey is very sweet and makes a great Pure Whey Ricotta, page 260.) Consolidate all of the curds on one side of the pot, removing as much of the remaining whey as you can. Gently press the curds together to release more whey. Divide the curds between three basket molds, pressing lightly to consolidate the curds in each mold.

6. Let the curd ripen in a warm spot. Rig up an incubator by pouring a few inches of warm water (90–100°F/32–38°C) into a pot with a rack in it. Put the molds onto the rack so they don't touch the water. Cover the pot and let the curd sit undisturbed for 5–6 hours.

7. After 5 hours, heat a pot of water to 175–185°F (79–85°C). Slice off a ¼-inch sliver of curd and submerge it into the hot water. Give it a few moments to heat up, then try to stretch it. If it stretches, it's ready; otherwise, let the curd rest for 15 minutes and test it again. If you're working with a pH meter, the curd should register 5.2–5.3 when ready to stretch.

8. Slice the curd into ¼-inch strips directly into a large heatproof bowl. Slowly pour the hot water down the interior side of the bowl, taking care not to pour directly on the curds, until the curds are submerged. Keep the curds moving to prevent sticking and promote even heating. When the curd begins to stretch, pour off the water and replace it with fresh hot water. Lift the curd out of the water with a wooden spoon, encouraging it to stretch and meld together into one single mass. Continue to work the curd with the spoon, folding it back on itself and stretching it as much as possible. If the curd gets stiff and difficult to stretch, add more hot water to reheat it.

9. As the surface of the curd becomes smooth, put on a pair of heatproof gloves and start stretching with your hands. Give the curd a few nice long stretches, folding it back on itself several times at the end of each stretch.

Recipe continues on next page

10. Fold the curd into a flat square. Begin to build the oval shape of the cheese by stuffing the edges of the square into the center of the underside of the square. Continue to build the cheese in this way, stuffing the edges into the center to create a large bag shape with a narrow neck. If the cheese gets stiff, reheat it in the hot water. You will end up with a large smooth oval ball with a topknot. Tie a piece of twine around the narrow neck and let the cheese hang for a few minutes to elongate into a pear shape. Then cut the twine and submerge the cheese in cold water, holding it with your hands to prevent it from settling on the bottom of the vessel.

11. When the surface of the cheese is cool, float it in the saturated brine for 4–6 hours, turning it once to make sure the cheese is evenly salted.

12. Tie a loop of twine around the topknot of the cheese. Hang the cheese at a temperature of 55°F (13°C) and at least 60% relative humidity to age for at least 6 weeks.

Ricotta Salata

Ricotta salata is a dry, salted ricotta that may be eaten at a young age as a sliceable dessert cheese or, as is much more common, aged to make a grating cheese. Traditionally, ricotta salata is made from ewe's-milk whey. Here is a variation you can enjoy at home.

1 recipe Whole-Milk Ricotta (page 82)

Cheese salt

Light Brine (page 46)

Yield: About ½ pound

1. Follow the recipe for Whole-Milk Ricotta through step 5.

2. Remove the ricotta from the bag, add 1 tablespoon of salt, and mix well.

3. Press the cheese into a ricotta mold set over a container to catch the whey, put a saucer or glass of water on the top, and press for 1 hour.

4. Unmold the cheese, turn it over, and put it back into the mold. Press for 12 hours.

5. Unmold the cheese and lightly sprinkle the surface with salt. Cover and refrigerate.

6. Turn the cheese and rub the entire surface with salt every day for 1 week. If any unwanted mold appears, gently rub it off with cheesecloth dampened in light brine. If the cheese becomes too soft with moisture, gently pat it dry, salt the surface again, and return it to the refrigerator.

7. Age the cheese in a sealed container in the refrigerator for 2–4 more weeks. Remove any whey/moisture left in the container. Wipe the surface with light brine if mold appears. For a grating version, age up to 2 months total.

Parmesan

Of all the many Italian cheeses, Parmesan is hands down the most beloved, thanks to the nutty taste and granular texture. No surprise that it comes from the food capital of Italy, Emilia-Romagna, and the same town that produces Prosciutto di Parma.

Parmesan is usually made from low-fat (2% butterfat) cow's milk and aged for at least 10 months to develop the highly prized sharpness. (Longer aging will give you a sharper flavor. I once aged a larger version of this cheese for 7 years, hanging from a rafter in the basement. It was incredible.) For even more piquancy, replace half the cow's milk with goat's milk, or add lipase as directed.

2 gallons 2% cow's milk

½ teaspoon calcium chloride diluted in ¼ cup cool, nonchlorinated water

1 packet direct-set thermophilic starter culture

¼ teaspoon lipase powder dissolved in ¼ cup cool, nonchlorinated water and allowed to sit for 20 minutes, for stronger flavor (optional)

½ teaspoon liquid rennet (or ½ rennet tablet) diluted in ¼ cup cool, nonchlorinated water

1 gallon Saturated Brine (page 46)

Light Brine (page 46)

1 teaspoon olive oil

Yield: About 2 pounds

1. Heat the milk to 90°F (32°C). Add the calcium chloride solution and stir well to combine. Sprinkle the starter over the surface of the milk, wait 2 minutes for the powder to rehydrate, then stir well. Add the lipase solution, if desired. Cover and let ripen for 30 minutes.

2. Make sure the milk's temperature is 90°F (32°C). Add the diluted rennet and stir gently with an up-and-down motion for 30 seconds. You may need to top-stir if using cream-line milk. Cover and let set at 90°F for 30 minutes, or until the curd gives a clean break.

3. Cut the curd into ¼-inch cubes.

4. Heat the curds to 100°F (38°C), raising the temperature 2 degrees every 5 minutes, stirring often.

5. Heat the curds to 124°F (51°C), raising the temperature 3 degrees every 5 minutes. Stir every 3 minutes to prevent the curds from matting. The curds will now be about the size of a grain of rice and they will squeak when chewed. Allow the curds to set for 5 minutes.

6. Carefully pour off the whey without losing any of the curd particles. Line a 2-pound cheese mold with cheesecloth. Pack the curds into the mold and press at 5 pounds of pressure for 15 minutes.

7. Remove the cheese from the mold, unwrap, flip, and rewrap. Press at 10 pounds for 30 minutes.

8. Remove the cheese from the mold, unwrap, flip, and rewrap. Press at 15 pounds for 2 hours.

9. Remove the cheese from the mold, line it with a fresh cloth, and put the cheese back into the mold. Press at 20 pounds of pressure for 12 hours.

10. Remove the cheese from the press. Peel away the cheesecloth. Soak the cheese in the saturated brine for 8 hours at 72°F (22°C).

11. Remove the cheese from the brine and pat dry. Age at 55°F (13°C) and 85% relative humidity for 10 months and up to 24 months. Turn the cheese daily for the first several weeks, and weekly after that. Remove any mold with a cloth dampened in light brine. After the cheese has aged for 2 months, rub the surface with olive oil to keep the rind and cheese from drying out.

ROMANO

As its name implies, Romano was first produced near Rome, and typically with ewe's milk. Now you will find it made with both cow's and goat's milk in regions of southern Italy and in Sardinia. Aged for between five and eight months, Romano makes a perfect table cheese. Longer aging (usually more than one year) produces a wonderfully piquant grating cheese. For a sharper-flavored Romano, replace half the cow's milk with goat's milk, or add lipase powder as directed.

2 gallons milk

6 ounces heavy cream

½ teaspoon calcium chloride diluted in ¼ cup cool, nonchlorinated water

½ packet direct-set thermophilic starter culture or 3.2 ounces prepared Bulgarian yogurt culture

¼ teaspoon lipase powder dissolved in ¼ cup cool, nonchlorinated water or ½ cup milk and allowed to sit for 20 minutes, for stronger flavor (optional)

½ teaspoon liquid rennet (or ½ rennet tablet) diluted in ¼ cup cool, nonchlorinated water

1 gallon Saturated Brine (page 46)

Cheese salt

Light Brine (page 46)

1–2 tablespoons olive oil

Yield: About 2 pounds

1. Heat the milk to 90°F (32°C). Add the cream. Add the calcium chloride solution and stir well to combine. Sprinkle the starter over the surface of the milk, wait 2 minutes for the powder to rehydrate, then stir well. (If you're using yogurt, simply stir it to eliminate any lumps and stir it into the milk.) Add the lipase solution, if desired. Cover and allow the milk to ripen for 10 minutes.

2. Add the diluted rennet, stirring gently with an up-and-down motion for 30 seconds. Cover and allow to set at 90°F (32°C) for 30 minutes, or until the curd gives a clean break.

3. Use a curd knife to cut the curd into 1-inch columns. Let rest for 3 minutes, then use a stainless-steel whisk to cut the curd into ¼-inch cubes. Stir gently every 3 minutes for 10 minutes.

4. Heat the curds to 116°F (47°C) over the course of 50 minutes, raising the temperature 2 degrees every 5 minutes at first, then gradually increasing to 1 degree per minute. Stirring gently every 3 minutes to prevent matting maintain the curds at 116°F for 30 minutes, or until they become firm enough that they retain their shape when squeezed. Let the curds settle to the bottom for 10 minutes undisturbed.

5. Drain off half of the whey.

6. Put the curd mass into a colander lined with butter muslin and drain for a few seconds. Leaving the curds in the cloth, place them into a 2-pound cheese mold with a follower and press the cheese at the following weights. With each increase of weight, unwrap the cheese, flip it, and rewrap it.

   - 5 pounds for 30 minutes
   - 10 pounds for 30 minutes
   - 15 pounds for 1 hour
   - 15 pounds for 1 hour
   - 15 pounds for 1 hour

7. Remove the cheese from the mold. Peel away the cheesecloth. Soak the cheese in the saturated brine for 8 hours at 50°F (13°C). Salt the exposed surface. After 4 hours, flip the cheese in the brine and salt the exposed surface.

8. Remove the cheese from the brine and pat dry. Air-dry for 2–3 days at 65–75% relative humidity.

9. Age the cheese at 55°F (13°C) and 75–85% relative humidity. Turn twice daily for one week, removing any mold with a cloth slightly dampened in light brine.

10. Once dried, lightly rub the cheese with the oil to keep the rind from drying out and continue aging for 6–24 months depending on the desired character.

Montasio

Legend has it that Montasio was first made in the seventeenth century in a monastery in northern Italy, and today the cheese is still only produced in Friuli-Venezia Giulia and East Veneto, the regions neighboring Venice. Montasio is enjoyed as a table cheese when aged for three months and as a grating cheese when made with skimmed milk for a drier texture and aged for one year or longer. To make Montasio with a sharper flavor, replace half the cow's milk with goat's milk.

2 gallons milk

½ teaspoons calcium chloride diluted in ¼ cup cool, nonchlorinated water

1 packet direct-set thermophilic starter culture

½ packet direct-set mesophilic starter culture

½ teaspoon liquid rennet (or ½ rennet tablet) diluted in ¼ cup cool, nonchlorinated water

1 gallon Saturated Brine (page 46)

Yield: About 2 pounds

1. Heat the milk to 88°F (31°C). Add the calcium chloride solution and stir well to combine. Sprinkle the starters over the surface of the milk, wait 2 minutes for the powders to rehydrate, then stir well. Cover and allow the milk to ripen for 60 minutes.

2. Make sure the milk's temperature is 88°F (31°C). Add the diluted rennet and stir gently with an up-and-down motion for 30 seconds. Cover and allow the curds to set at 88°F for 30 minutes, or until the curd gives a clean break.

3. Use a curd knife to cut the curd into 1-inch columns. Let rest for 3 minutes, then use a stainless-steel whisk to cut the curd into ¼-inch cubes. Stir gently every 3 minutes for 10 minutes.

4. Heat the curds to 102°F (39°C), raising the temperature 2 degrees every 5 minutes and stirring every 3 minutes to keep the curds from matting. Maintain the temperature at 102°F for 60 minutes and continue to stir.

5. Remove the whey until the liquid level is at the level of the curds. Add 125°F (52°C) water over 5 minutes until the curd-whey mixture reaches 110°F (43°C). Maintain that temperature for 10 minutes, stirring gently every 3 minutes. Drain off the whey.

6. Line a 2-pound cheese mold with cheesecloth. Quickly ladle the curds into the mold. Press at 5 pounds of pressure for 15 minutes.

7. Remove the cheese from the mold and peel away the cheesecloth. Turn over the cheese, rewrap it, and press at 5 pounds of pressure for 30 minutes.

8. Unwrap, flip, rewrap, and press at 10 pounds of pressure for 12 hours.

9. Unwrap the cheese and soak in saturated brine at 72°F (22°C) for 6 hours, turning halfway through. Remove the cheese from the brine and pat dry.

10. Age the cheese at 55–60°F (13–16°C) and about 85% relative humidity for 2–12 months.

IBORES,
page 190

SPANISH-STYLE CHEESES

Ask anyone to name a Spanish cheese and they will most likely answer with "Manchego." This limited knowledge of Spanish cheese is a direct result of Franco's long rule of Spain in the twentieth century, during which many regional cheeses were outlawed in favor of those made in large commercial cheese factories. Fortunately, many rural cheese makers continued making and selling cheese locally, but none made it out of Spain until fairly recently.

The one positive aspect of this limited production of Spanish cheeses is that it preserved some of the heirloom cheeses from becoming overly industrialized and losing their traditional character. When the ban on making these cheeses was lifted, those who knew the recipes came forth and revived the traditional processes as they best remembered them.

Much of the land in Spain is so hot and dry that only sheep and goats can coax any nutrition from the land. From the sweet ewe's milk used for Manchego to the full-fat goat's milk transformed into the smoked paprika–rubbed Ibores, Spanish cheeses take on the character of the wild, often communal pastureland that provides such a perfect daily meal for sheep and goats.

MANCHEGO,
page 188

MANCHEGO–STYLE CHEESE

This rich, mellow cheese is a specialty of Spain, where it was originally made near the plains of Toledo using the milk of Manchego sheep. It is available in four stages of ripeness: Manchego fresco is aged for 5 days or fewer; Manchego curado is aged for 3–12 weeks; Manchego viejo is aged for 3–12 months; and Manchego aceite is aged in olive oil for more than 1 year. Today, Manchego is made almost entirely from cow's milk.

2 gallons milk

½ teaspoon calcium chloride diluted in ¼ cup cool, nonchlorinated water

½ packet direct-set mesophilic starter culture

½ packet direct-set thermophilic starter culture

¼ teaspoon lipase powder dissolved in ¼ cup of water and allowed to sit for 20 minutes before using, for stronger flavor (optional)

½ teaspoon liquid rennet (or ½ rennet tablet) diluted in ¼ cup cool, nonchlorinated water

1 gallon Saturated Brine (page 46)

Cheese salt

Olive oil

Yield: 2 pounds

1. Heat the milk to 72°F (22°C). Add the calcium chloride solution and stir well to combine. Sprinkle both starters over the surface of the milk, wait 2 minutes for the powders to rehydrate, then stir well. Cover and allow the milk to ripen for 15 minutes at 72°F. Heat the mixture to 86°F (30°C) and let set at 86°F for 30 minutes.

2. Add the lipase solution, if desired. Add the diluted rennet and stir gently with an up-and-down motion for 30 seconds. Cover and let set at 86°F (30°C) for 30 minutes, or until the curd gives a clean break.

3. Cut the curd vertically into ¾-inch columns and let sit for 5 minutes.

4. Cut the curds into rice-size pieces by slowly stirring them with a stainless-steel whisk.

5. Heat the curds to 102°F (39°C) by increasing the temperature 2 degrees every 5 minutes over the next 45 minutes. Stir gently every 3 minutes to prevent the curds from matting.

6. Let the curds sit for 5 minutes, then pour off the whey.

7. Ladle the curds into a 2-pound cheese mold lined with cheesecloth. Press at 15 pounds of pressure for 15 minutes.

8. Remove the cheese from the mold and gently peel away the cheesecloth. Turn over the cheese, rewrap it, and press at 15 pounds of pressure for 15 minutes.

9. Unwrap, flip, rewrap, and press at 15 pounds of pressure for 15 minutes.

10. Unwrap, flip, rewrap, and press at 30 pounds of pressure for 6 hours.

11. Remove the cheese from the mold and place it in the saturated brine at 55°F (13°C) for 6 hours. Sprinkle the exposed surface with salt. Flip the cheese halfway through and sprinkle the exposed surface with salt.

12. Remove the cheese from the brine and pat dry with a piece of butter muslin. Coat lightly with oil to keep the cheese from drying out. Age at 55°F (13°C), turning daily. Age for any length of time (see headnote), depending on the type of cheese you want, periodically removing unwanted mold with cheesecloth dampened with brine.

|BORES

Ibores is historically a full-fat goat's-milk cheese, originally from the Extremadura region of Spain. Traditionally, the cheese has been made with no added culture, ripening only with the natural bacteria of the milk and milking environment. It has a striking orange-red rind, which comes from a paste of smoked paprika and olive oil. The flavor and aroma of this smoky paprika blends well with the bright supple character of the cheese. Over the time that the cheese spends in the aging room, this flavor permeates the cheese.

3 gallons milk

¼ teaspoon calcium chloride diluted in ¼ cup of cool, nonchlorinated water

1/16 teaspoon MA4002 mesophilic and thermophilic blend starter culture

¼ teaspoon liquid rennet (or ¼ rennet tablet) diluted in ¼ cup cool, nonchlorinated water

1 gallon Saturated Brine (page 46)

Cheese salt

2 tablespoons smoked paprika

Olive oil

Yield: 3 pounds

1. Slowly heat the milk to 86°F (30°C). Add the calcium chloride solution and stir well to combine. Sprinkle the starter over the surface of the milk, wait 2 minutes for the powder to rehydrate, then stir well. Cover the pot and let the milk ripen for 1 hour at 86°F.

2. Add the diluted rennet and stir gently with an up-and-down motion for 30 seconds. Cover and let set undisturbed for 1 hour at 86°F (30°C), or until the curd gives a clean break.

3. Cut the curd vertically to make 2-inch-square columns. Let the curds sit and release whey for a few minutes, then cut them into ¼-inch cubes with a wire whisk. Let the curds rest for 10 minutes, stirring once or twice to prevent them from sticking together. The curds will be very soft; treat them gently.

4. Over the next 30 minutes, stir gently every 3 minutes while keeping the curds at 86°F (30°C). Remove the pot from the heat and let the curds settle in the whey. Remove half of the whey from the pot so that the liquid level is just to the curd level. Give the curds a good stir. Line a 2- to 4-pound cheese mold with butter muslin.

5. Use a ladle to transfer the curds and whey to the cheese mold. Fold the butter muslin over the curds, and place a follower on top. Press for 1 hour at 25 pounds.

6. Unwrap the cheese, flip, rewrap, and return to the press. Increase the weight to 50 pounds and press for 90 minutes. Unwrap, flip, rewrap, and press at 75 pounds for 4 hours. The final cheese will have a smooth surface.

7. Unmold the cheese and place it in the saturated brine for 3 hours. Sprinkle the exposed surface with salt. Flip the cheese halfway through the soak time, and sprinkle the other side with salt. Remove from the brine, turn, and wipe the surface of the cheese, and allow it to dry at room temperature for 1–2 days. The surface will get a bit darker over this time and a white powder may appear. Brush the surface of the cheese to remove any mold or surface powder.

8. Put the paprika in a small bowl and add just enough oil to create a thick paste. Rub the entire surface of the cheese with the paprika mixture. The oil will absorb into the cheese within a day, leaving a fine layer of dry paprika.

9. Age the cheese at 52–56°F (11–13°C) and 80–85% relative humidity for 4–6 weeks, turning twice a week. If mold grows on the cheese at any point during the aging process, simply brush it off and repeat the paprika rub.

São Jorge

Island cheese makers in the Azores combine the morning milk with the evening milk to get the cheese started by dark. They make the tangy semi-hard cheese all night, using some of the whey as the starter for the next batch. As always, the flavor of the cheese has everything to do with the traditions of the place. That said, this recipe will help you make delicious São Jorge at any time of day, right in your own kitchen.

2 gallons milk

½ teaspoon calcium chloride diluted in ¼ cup cool, nonchlorinated water

⅛ teaspoon MA4002 mesophilic and thermophilic blend starter culture

¼ teaspoon liquid rennet (or ¼ rennet tablet) diluted in ¼ cup cool, nonchlorinated water

1 tablespoon cheese salt

Yield: 1¾ pounds

1. Slowly heat the milk to 87°F (31°C). Add the calcium chloride solution and stir well to combine. Sprinkle the starter over the surface of the milk, wait 2 minutes for the powder to rehydrate, then stir well. Cover the pot and let the milk ripen at 87°F for 40 minutes.

2. Add the diluted rennet and stir gently with an up-and-down motion for 30 seconds. Cover and let set undisturbed at 87°F (31°C) for 1 hour, or until the curd gives a clean break.

3. Cut the curd into ⅜-inch cubes and let it rest for 5 minutes to firm up. Then stir gently every 3 minutes for 10 minutes at 87°F (31°C).

4. Over the next 30–45 minutes, stir gently, slowly heating the curds to 97°F (36°C), increasing the heat by 2–3 degrees every 5 minutes. The final curd will be firm throughout and have a moderate resistance when pressed between the fingers. If the curd is too soft, continue to stir and cook at 97°F for up to an hour. Remove the pot from the heat, and let the curd settle in the whey.

5. Remove enough whey from the pot so the liquid level is just above the curd level. Use a ladle to transfer the curd and remaining whey to a colander. Quickly wash out your pot, and transform it into an incubator to keep the curd warm as it drains: Pour a few inches of 95°F (35°C) water into the pot, and fit the pot with a rack or steamer insert that will support the

colander close to but not in the water. Put the colander on the rack, cover the pot, and let the curd drain for about 3 hours, keeping the curd temperature between 75 and 80°F (24–27°C). Lift the lid to stir the curd very gently every 3 minutes for the first 20 minutes and then every 30 minutes for the remaining time.

6. Remove the colander from the pot. Sprinkle the salt over the curd, gently tossing with your hands to distribute the salt evenly. Gather the corners of the cloth and transfer the curd in the cloth to a small hard-cheese mold, using your hands to gently reshape the curd to fit into the mold.

7. Press the cheese at 20 pounds of pressure for 30 minutes, at 40 pounds for 1 hour, at 50 pounds for 4 hours, and finally at 75 pounds for 2 days. Each time you increase the weight, remove the cheese from the mold and unwrap, flip, rewrap, and return it to the mold. Try to keep the curd at 72°F (22°C) throughout the pressing stage. When you unmold the cheese after pressing, the surface will be smooth.

8. Age the cheese at 58–65°F (14–18°C) and 70–80% relative humidity for 3–4 weeks. Dry-brush the surface of the cheese every day after 3–5 days, turning it each time. If the spaces between the curds begin to shrink and crack during this time, increase the humidity.

9. Transfer the cheese to an aging space of 52–56°F (11–13°C) and 82–88% relative humidity. Age for 3–7 months, turning twice weekly and dry-brushing the surface if any mold appears.

EMMENTAL,
page 199

GRUYÈRE, *page 202*

JARLSBERG-STYLE,
page 196

OTHER EUROPEAN—STYLE CHEESES

Europe might not be a sprawling continent, but the richness in the diversity of cheese that the small area has created is a wonder. It's no surprise when we take into account the varying landscapes, cuisines, the plants the animals feed on, and the animals themselves that hold an essential place in the culture of each country. From creamy Norwegian Jarlsberg to the iconic Swiss Gruyère, from tangy Dutch Gouda to dill-flecked Danish Havarti — these cheeses provide a tour of Europe, bite by delicious bite.

JARLSBERG–STYLE CHEESE

Jarlsberg is a Norwegian cheese, sweeter and milder than a Swiss Alpine cheese, with a creamy flavor and dense texture reminiscent of Gouda. The actual process is a secret recipe, created with protected cultures that we can't replicate. This recipe, however, creates a decent copy, complete with the trademark holes and rich flavor.

4 gallons milk

1 teaspoon calcium chloride diluted in ¼ cup cool, nonchlorinated water

¼ teaspoon MM100 mesophilic starter culture, Aroma B, or Flora Danica starter culture*

¹⁄₁₆ teaspoon *Propionic shermanii* powder

½ teaspoon liquid rennet (or ½ rennet tablet) diluted in ¼ cup cool, nonchlorinated water

1 gallon Saturated Brine (page 46)

Cheese salt

Light Brine (page 46)

*Flora Danica may be used as either a direct-set or a reculturable starter (see page 15).

Yield: 4 pounds

1. Slowly heat the milk to 98°F (37°C). Add the calcium chloride solution and stir well to combine. Sprinkle the starter and *P. shermanii* over the surface of the milk, wait 2 minutes for the powders to rehydrate, then stir well. Cover the pot and let the milk ripen for 45 minutes at 98°F.

2. Add the diluted rennet and stir gently with an up-and-down motion for 30 seconds. Cover and let set undisturbed until the curd gives a clean break, about 30 minutes.

3. Cut the curd vertically into 1-inch square columns. Let the curd rest as the whey rises in the cuts for about 3 minutes. Cut the curds into ¼-inch cubes with a whisk. Stir the curds gently for 20 minutes, then let them settle to the bottom of the pot. Remove one-third of the whey from the pot, scooping it into a measuring cup before discarding, as for the next step you'll need to know the quantity of whey removed. Give the curds a good stir.

4. Begin to gradually add 140°F (60°C) water to the pot, stirring to keep the curds from sticking together. The quantity of water you'll add will be half of the quantity of removed whey. Add the warm water over the course of 20–30 minutes. At the end of this time, the temperature of the whey in the pot will be 102°F (39°C).

5. Hold the curd at 102°F (39°C), continuing to stir until the curd is cooked through and resistant to the touch, 30–45 minutes. Let the curds settle to the bottom of the pot. Remove the whey until the liquid level is 2 inches above the curds.

6. Gather the curd in the pot into a square yard of butter muslin. Take the cloth by the corners, and quickly put the bundle into a large hard-cheese mold. Then submerge the whole mold back into the whey. Gently press on the curd with your hands to push it deeper into the mold. Top the mold with a follower and add 8 pounds of weight. Let the cheese sit for 15 minutes in the whey.

7. Taking the mold out of the whey, remove the cheese, unwrap, flip, and rewrap. Set the mold on a tray to continue draining at 80–85°F (27–30°C), using 8 pounds of pressure for 30 minutes.

Recipe continues on next page

8. Remove the cheese from the mold again, unwrap, flip, and rewrap. Repeat this process every hour for 4 hours. Take care to pull the cloth tight, as any wrinkles will show up on the surface of the cheese. After 4 hours, remove the weight from the cheese and let it rest in the mold for 2–3 hours. Transfer the cheese to a place at 61–64°F/16–18°C to cool for 8–10 hours.

9. Unmold the cheese, remove the wrapping, and place the cheese in the saturated brine for 8 hours. Sprinkle the exposed surface with salt. Flip the cheese halfway through the soak time and sprinkle the other side with salt. Wipe the cheese and let it air-dry for 2 days. When it is totally dry, brush the surface of the cheese and wipe it with light brine.

10. Age the cheese at 52–54°F (11–12°C) and 92–95% relative humidity for 7–10 days. Brush the surface and wipe it with light brine if mold develops. Transfer the cheese to a warmer (68–72°F/20–22°C), enclosed space with 92–95% relative humidity for 4–5 weeks, turning the cheese daily and wiping it with light brine if mold develops. A large plastic container or cake carrier with a lid works well. The cheese will swell over this time.

11. Return the cheese to the colder (52–54°F/11–12°C) space and 92–95% relative humidity. Continue to age for at least 3 months, turning twice weekly, brushing the surface and wiping with light brine if any mold develops. The cheese will be ready at 3 months, and can be aged longer to develop a more complex flavor.

EMMENTAL

Also known as Emmentaler, this most famous Swiss cheese owes its name to the Emmental, a valley near Bern, Switzerland.

2 gallons milk

½ teaspoon calcium chloride diluted in ¼ cup cool, unchlorinated water

1 packet direct-set thermophilic starter culture

⅛ teaspoon *Propionic shermanii* powder

½ teaspoon liquid rennet (or ½ rennet tablet) diluted in ¼ cup cool, nonchlorinated water

1 gallon Saturated Brine (page 46)

Yield: 2 pounds

1. Heat the milk to 90°F (32°C). Add the calcium chloride solution and stir well. Sprinkle the starter over the surface of the milk, wait 2 minutes for the powder to rehydrate, then stir well to combine.

2. Remove ¼ cup of the milk and add the *P. shermanii* to it. Mix thoroughly, making sure the powder is dissolved. Add it to the milk in the pot and stir. Cover and allow the milk to ripen for 10 minutes.

3. Make sure the milk's temperature is 90°F (32°C), then add the diluted rennet and stir gently with an up-and-down motion for 30 seconds. If using cream-line cow's milk, top-stir for 30 seconds longer. Cover and let set at 90°F for 30 minutes, or until the curd gives a clean break.

4. Cut the curd into ¼-inch cubes using a curd knife and a stainless-steel whisk. Keeping the curd temperature at 90°F (32°C), gently stir the curds every 3 minutes for 20 minutes to prevent matting.

5. Over the next 30 minutes, heat the curds by 2 degrees every 5 minutes, until the temperature reaches 100°F (38°C).

6. Over the next 30 minutes, heat the curds 1 degree every 2 minutes, until the temperature reaches 114°F (46°C). Maintain this temperature for 30 minutes, stirring every 3 minutes to prevent matting, until they reach the proper break. To test for the proper break, gather a handful of curds. Rub the mass gently between your palms. If the ball of curds readily breaks apart into individual curd particles, the curds are sufficiently cooked. Let the curds set for 5 minutes.

7. Follow steps 7–12 for Traditional Swiss-Style (see page 200).

TRADITIONAL SWISS–STYLE

A delicious mountain cheese, made in the spring when the cows are out grazing on beautiful lush grass. The longer it ages, the more flavor is developed. You can add caraway seeds for a delicious variation.

2 gallons milk

½ teaspoon calcium chloride diluted in ¼ cup cool, nonchlorinated water

1 packet direct-set thermophilic starter culture

⅛ teaspoon *Propionic shermanii* powder

½ teaspoon liquid rennet (or ½ rennet tablet) diluted in ¼ cup cool, nonchlorinated water

1 gallon Saturated Brine (page 46)

Cheese salt

Light Brine (page 46)

Yield: 2 pounds

1. Heat the milk to 90°F (32°C). Add the calcium chloride solution and stir well to combine. Sprinkle the starter over the surface of the milk, wait 2 minutes for the powder to rehydrate, then stir well to combine.

2. Remove ¼ cup of milk and add the *P. shermanii* to it. Mix thoroughly to dissolve the powder. Add the mixture to the rest of the milk and stir. Cover and allow the milk to ripen for 10 minutes.

3. Make sure the milk is still at 90°F (32°C). Add the diluted rennet and stir gently with an up-and-down motion for 30 seconds. If using cream-line cow's milk, top-stir for 30 seconds longer. Cover and let set at 90°F for 30 minutes, or until the curd gives a clean break.

4. Using a curd knife and a stainless-steel whisk, cut the curd into ¼-inch cubes.

5. Keeping the curd temperature at 90°F (32°C), gently stir the curds for 40 minutes, stirring every 3 minutes to keep the curds from matting.

6. Heat the curds by 1 degree every minute until the temperature is 120°F (49°C); this will take 30 minutes. Maintain the temperature at 120°F for 30 minutes, stirring every 3 minutes to prevent matting. The curds must be cooked until they reach a stage called the proper break. To test for this, gather a handful of curds and rub it gently between your palms. If the ball readily breaks apart into individual particles, the curds are sufficiently cooked. If they are not sufficiently cooked, they will be too soft to hold together. Let the curds set for 5 minutes.

7. Pour off the whey.

8. Line a 2-pound mold with cheesecloth and place it in the sink or over a large pot. Quickly ladle the curds into the mold. You do not want the curds to cool. Top the mold with a follower, and press the cheese at the following weights. With each increase of weight, unwrap the cheese, flip it, and rewrap it.

   - 8–10 pounds for 15 minutes

   - 14 pounds for 30 minutes

   - 14 pounds again for 2 hours

   - 15 pounds for 12 hours

9. Remove the cheese from the mold, peel away the cheesecloth, and place the cheese in the saturated brine. Sprinkle a pinch of salt on the surface of the floating cheese. Refrigerate in the brine and let the cheese soak for 12 hours.

10. Remove the cheese from the brine and pat dry with sanitized butter muslin. Place on a clean cheese board and store at 50–55°F (10–13°C) and 85% relative humidity. Turn daily for 1 week, wiping it with clean cheesecloth dampened in light brine. Do not wet the cheese.

11. Place the cheese in a warm, humid room, with a temperature between 65 and 70°F (18–21°C) for 2–3 weeks, until eye formation is noticeable. Turn it daily and wipe it with a cheesecloth dampened in light brine. (The cheese will swell somewhat and become slightly rounded.)

12. Age the cheese at 45°F (7°C) and 80% relative humidity for 3–6 months. Turn the cheese several times a week. Dry-brush any surface mold. A reddish coloration on the surface of the cheese is normal and should not be removed.

FLAVOR VARIATION

Boil ½–2 tablespoons caraway seeds in ½ cup water for 15 minutes, adding more water to cover the seeds as needed. Strain the flavored water into a small bowl and let cool; reserve the boiled seeds. If desired, you can add the water to the milk at the beginning of the recipe.

Add the seeds after you drain off the whey. Hold back 2 cups of plain curds, then gently mix the seeds into the remaining curd. Blend thoroughly and quickly. You do not want the curds to cool before they are placed in the cheese mold.

When filling the mold, leave a top and bottom layer of plain curd so seeds do not poke out, which can create air pockets for mold to grow.

GRUYÈRE

This is the classic mountain cheese of France and Switzerland, differentiated from Emmental cheese by its tiny or nonexistent holes. This cheese relies on a high-temperature scalding of the curd to prepare it for a long aging period.

4 gallons milk

1 teaspoon calcium chloride diluted in ¼ cup cool, nonchlorinated water

1 packet C201 thermophilic starter culture

¹⁄₃₂ teaspoon *Propionic shermanii* powder

½ teaspoon liquid rennet (or ½ rennet tablet) diluted in ¼ cup cool, nonchlorinated water

1 gallon Saturated Brine (page 46)

Cheese salt

Light Brine (page 46)

Yield: 4 pounds

1. Slowly heat the milk to 90°F (32°C). Add the calcium chloride solution and stir well to combine. Sprinkle the starter and the *P. shermanii* over the surface of the milk, wait 2 minutes for the powders to rehydrate, then stir well. Cover the pot and let the milk ripen for 1 hour at 90°F.

2. Add the diluted rennet and stir gently with an up-and-down motion for 30 seconds. Cover and let set at 90°F (32°C) until the curd gives a clean break, about 30 minutes.

3. Cut the curd vertically into 1-inch columns, and let them rest for 5 minutes. Using a wire whisk, cut the curds into ¼-inch cubes. Gently stir the curds for 1 minute.

4. Over the next 30 minutes, heat the curds 1 degree every minute to 114°F (46°C), while slowly stirring. Continue to stir the curds at 114°F until they are firm. The final result will be an elastic curd, fully cooked through and resistant to the touch.

5. Remove the pot from the heat and let the curds settle in the whey for 5 minutes. Remove the whey from the pot so that the liquid level is just to the curd level. Put a plate over the curds and place an 8- to 12-pound weight on the plate. Let the curds rest until they come together in a large mass.

6. Gather the curds in butter muslin and transfer them to a large Tomme mold. Press the cheese at 25 pounds of pressure for 4 hours. Remove the cheese from the mold, unwrap, flip, rewrap, and return to the mold. Press at 50 pounds for 6 hours. Unwrap, flip, rewrap, and press for 10 hours at 75 pounds.

7. Remove the cheese from the mold, unwrap, and place in the saturated brine for 12 hours. Sprinkle the exposed surface with salt. Halfway through the soak time, flip the cheese and sprinkle the other side with salt. Remove the cheese from the brine and pat dry with sanitized cheesecloth.

8. Age the cheese at 54–58°F (12–14°C) and 85% relative humidity until a rind develops, 30–40 days. Rub the surface with salt several times over the first 2 days of aging, and wipe away any mold that develops on the rind with light brine. Age the cheese for 8–14 months or more, turning it and washing the rind with light brine 2 or 3 times per week.

GOUDA, *page 206*

HAVARTI, *page 208*

Gouda

Hailing from the Dutch town of Gouda, near Rotterdam, this popular cheese is a washed-curd, semi-hard cheese with a smooth texture and a deliciously tangy taste. It looks particularly distinctive when covered with red cheese wax, though that is not done in Holland. Goat's-milk Gouda is especially delicious. For flavor variations, see the instructions on page 144 for the Stirred-Curd Cheddar variations.

2 gallons milk

½ tsp calcium chloride diluted in ½ cup cool, nonchlorinated water

1 packet direct-set mesophilic starter culture

½ teaspoon liquid rennet (or ½ rennet tablet) diluted in ¼ cup cool, nonchlorinated water

1 gallon Saturated Brine (page 46)

Cheese wax (optional)

Olive oil (optional)

Yield: 2 pounds

1. Heat the milk to 90°F (32°C). Add the calcium chloride solution and stir well to combine. Sprinkle the starter over the surface of the milk, wait 2 minutes for the powder to rehydrate, then stir well. Cover and allow the milk to ripen for 30 minutes.

2. Add the diluted rennet and stir gently with an up-and-down motion for 30 seconds. If using cream-line cow's milk, top-stir for 30 seconds longer. Cover and let the milk set at 90°F (32°C) for 45 minutes, or until it gives a clean break.

3. Cut the curd into ½-inch cubes. Let them set for 10 minutes.

4. Drain off one-third of the whey. Stirring continuously, slowly add just enough 175°F (79°C) water to raise the temperature of the curd to 92°F (33°C).

5. Let the curd settle again for 10 minutes. Drain off the whey so the liquid level is at the level of the curd.

6. Once again, while stirring, slowly add enough 175°F (79°C) water to bring the temperature of the curd to 100°F (38°C). Keep the curd at 100°F for 15 minutes, stirring gently every 3 minutes to keep the curds from matting.

7. Allow the curds to rest for 30 minutes. Pour off the remaining whey.

8. Quickly ladle the warm curds into a 2-pound cheese mold lined with cheesecloth, breaking them as little as possible. Press at 20 pounds of pressure for 20 minutes.

9. Remove the cheese from the mold and gently peel away the cheesecloth. Turn over the cheese, re-dress it, and press at 40 pounds of pressure for 20 minutes.

10. Unwrap, flip, rewrap, and press at 50 pounds of pressure for 12–16 hours. Remove from the press.

11. Soak the cheese in the saturated brine solution for 8 hours.

12. Remove the cheese from the brine and pat dry with sanitized cheesecloth. Air-dry the cheese at 50°F (10°C) for 3 weeks, turning daily.

13. Wax the cheese (see page 50) for a very moist cheese, or oil it and leave it to dry naturally and develop a natural rind for a drier cheese.

14. Age the cheese at 50°F (10°C) for 3–4 months, turning it 3 or 4 times a week. For a real treat, keep one around for 6–9 months.

TROUBLESHOOTING

If, after removing your cheese from the mold, it has no closed rind (the curds have not properly knit together), dip it in 100°F (38°C) water and press for 30 minutes.

ABOUT WASHED-CURD CHEESES

Gouda and Colby (page 212) are considered "washed-curd" cheeses. During cooking, the whey is removed from the pot and replaced with water, which washes the milk sugar, or lactose, from the curd and lowers the acid level to avoid souring the cheese. This process gives these cheeses their typical smooth texture and mild flavor, and also makes them ready for eating in just 12 weeks.

Havarti

Havarti is a simple, washed-rind cheese with a buttery aroma and a sweet taste, named after the farm in Denmark where it was first made. It is typically aged for three months, but if it's aged longer, the cheese becomes saltier and tastes like hazelnuts (a bonus in my book). Take note: When left at room temperature the cheese tends to soften quickly.

4 gallons milk

1 teaspoon calcium chloride diluted in ¼ cup cool, nonchlorinated water

1 packet C101 direct-set mesophilic starter culture

½ teaspoon liquid rennet (or ½ rennet tablet) diluted in ¼ cup cool, nonchlorinated water

Cheese salt

1 gallon Saturated Brine (page 46)

Light Brine (page 46)

Yield: 4 pounds

1. Slowly heat the milk to 86°F (30°C). Add the calcium chloride solution and stir well to combine. Sprinkle the starter over the surface of the milk, wait 2 minutes for the powder to rehydrate, then stir well. Cover the pot and let the milk ripen for 45 minutes at 86°F.

2. Add the diluted rennet and gently stir with an up-and-down motion for 30 seconds. Cover and let set undisturbed at 86°F (30°C) for 35 minutes, or until the curd gives a clean break.

3. Cut the curd into ⅜-inch cubes (a wide whisk works well). Let them rest for 5 minutes, then gently stir for 15 minutes. Remove one-third of the whey from the pot, and stir for an additional 15 minutes.

4. Heat a gallon of water to 130°F (54°C). Slowly add the hot water to the curds, pouring and stirring until the curds reach between 95 and 100°F (35–38°C). The higher the temperature, the drier the cheese will be. Add 2 tablespoons of salt to the pot and stir for an additional 20 minutes.

5. Transfer the curds to a large colander lined with butter muslin to drain for 10 minutes.

6. Transfer the drained curds into a large Tomme mold lined with butter muslin. Add 8 pounds of weight, and press for 15–20 minutes. Unwrap the cheese, flip it, and rewrap it. Increase the weight to 16 pounds and press for 2 hours, unwrapping, flipping, and rewrapping the cheese every 30 minutes.

7. Remove the cloth from the cheese and put it back into the mold. Submerge the mold in a bowl of 65°F (18°C) water for 8–10 hours.

8. Unmold the cheese and place it in the saturated brine for 5–6 hours. Sprinkle the exposed surface with salt. Flip the cheese halfway through the soak time and sprinkle the other side with salt. Place the cheese on a draining mat and let it sit at room temperature until dry to the touch, 1–3 days.

9. Age the cheese at 59°F (15°C) and 90% relative humidity for 5 weeks for a young cheese or 10–14 weeks for a mature cheese. Turn the cheese daily through the aging process, and wipe it down with light brine every 2–3 days.

10. After the initial aging is complete, store for one more week at 54°F (12°C) and 80% relative humidity.

VARIATION

Dill and other herbs are traditional additions to Havarti. To add herbs, start at step 6. When ladling the curd into the mold, build a ½-inch layer of curd at the bottom. Then add a thin layer of the herbs of your choice, leaving a ½-inch band of plain curd around the edge. (This will help the finished cheese encapsulate the herbs.) Continue to alternate layers of curd and herbs, completing the process with a layer of curd. See page 73 for more about adding herbs.

DRY JACK, *page 216*

BABY SWISS, *page 220*

AMERICAN-STYLE CHEESES

Like most Americans, American-style cheeses are immigrants. Each recipe can be traced back to a country of origin, often European. But in the journey across the ocean and the process of integration into a new home, these cheeses have become entirely distinct from their cheese of origin, and truly American.

Monterey Jack hails from Monterey, California, and many say the name refers to the "jack" device used to press the cheese. The method came from Spain, and the first family in Monterey to create the cheese had carried the method with them along with the possessions that would help start their new American life.

Baby Swiss has its own story, originating with Swiss cheese makers who, continuing to hone their craft in their new home, worked to create a milder, smaller version of Emmentaler more suited to the American palate. The result was a uniquely American cheese — buttery, nutty, and versatile.

Colby

A type of cheddar in which the curds are washed during the cooking stage, Colby is an American cheese named after the township in southern Wisconsin where it was first made. It has a more open texture than cheddar, contains more moisture, has a pleasantly mild flavor, and can be aged from two to three months.

2 gallons milk

½ teaspoon calcium chloride diluted in ¼ cup cool, nonchlorinated water

½ packet direct-set mesophilic starter culture

¼ teaspoon annatto cheese coloring diluted in ¼ cup water

½ teaspoon liquid rennet (or ½ rennet tablet) diluted in ¼ cup cool, nonchlorinated water

1 gallon Saturated Brine (page 46)

2 tablespoons cheese salt

Cheese wax (optional)

Yield: 2 pounds

1. Heat the milk to 86°F (30°C). Add the calcium chloride solution and stir well to combine. Sprinkle the starter over the surface of the milk, wait 2 minutes for the powder to rehydrate, then stir well. Cover and allow the milk to ripen for 1 hour.

2. Add the diluted annatto coloring and stir. (You will not be able to see the color change at this point because of the water content.)

3. Make sure the milk's temperature is 86°F (30°C). Add the diluted rennet and stir gently with an up-and-down motion for 30 seconds. If using cream-line cow's milk, top-stir for 3 minutes longer. Cover and let set for 30 minutes, or until the curd gives a clean break.

4. Slowly cut the curd into ½-inch cubes, handling the curd very gently to avoid shattering it.

5. Heat the curds to 102°F (39°C) by raising the temperature 2 degrees every 5 minutes over the next 30 minutes, stirring gently every 3 minutes to prevent matting. Let the curds settle at the bottom of the pot. The finished curds will have a moderate resistance when pressed between your fingers. If the curds are still soft, hold the temperature for 15–30 minutes longer, stirring every 3 minutes to prevent matting. Let the curds rest for 5 minutes.

6. Remove whey until the liquid level is at the level of the curds. Then, while gently stirring, add 75°F (24°C) water until the temperature comes down to 90°F (32°C). Once cooled, let the curds settle at the bottom for 3 minutes. Once settled, again remove the liquid down to the level of the curds. Stir in 60°F (15°C) water until the temperature cools to 75°F (24°C). Allow to sit for 15 minutes, stirring gently every 3 minutes to prevent matting. (The temperature of the additional water controls the moisture content of the cheese.)

7. Transfer the curds into a cheesecloth-lined colander and drain off any remaining whey.

8. Place the curds in a small hard-cheese mold lined with cloth. Once packed, pull the muslin tight. Top the mold with a follower, and press the cheese at the following weights. With each increase of weight, unwrap the cheese, flip it, and rewrap it.

   - 10 pounds for 15 minutes
   - 20 pounds for 30 minutes
   - 40 pounds for 90 minutes
   - 50 pounds for 8 hours

9. Unmold the cheese and place it in the saturated brine for 8 hours. Sprinkle the exposed surface with 2 teaspoons of salt. After 4 hours, flip the cheese and salt again. When done remove from the brine, wipe the cheese, and air-dry for 1–2 days. Flip twice daily.

10. Oil or wax the cheese (see page 50) and age at 52–56°F (11–13°C) and 80–85% relative humidity for 4–6 weeks, flipping 3 times weekly.

Monterey Jack

Created in 1892 in Monterey County, California, this cheese may be made with skim, part-skim, or whole milk. If you use whole milk, the cheese will be semi-soft; if you use skim milk, it will be hard enough to grate.

2 gallons milk

½ teaspoon calcium chloride diluted in ¼ cup cool, non-chlorinated water

1 packet direct-set mesophilic starter culture

½ teaspoon liquid rennet (or ½ rennet tablet) diluted in ¼ cup cool, nonchlorinated water

1 tablespoon cheese salt

Cheese wax (optional)

Yield: 1 pound

1. Heat the milk to 90°F (32°C). Add the calcium chloride solution and stir well to combine. Sprinkle the starter over the surface of the milk, wait 2 minutes for the powder to rehydrate, then stir well. Cover and allow the milk to ripen at 90°F for 30 minutes.

2. Add the diluted rennet and stir gently with an up-and-down motion for 30 seconds. Cover and let set for 30–45 minutes at 90°F (32°C), or until the curd gives a clean break.

3. Cut the curd into ¼-inch cubes and let set for 40 minutes.

4. Heat the curds to 100°F (38°C), increasing the temperature by 2 degrees every 5 minutes (this will take about 35 minutes). Stir gently every 3 minutes to keep the curds from matting.

5. Maintain the curds at 100°F (38°C) for 30 minutes, stirring gently every 3 minutes to keep the curds from matting. Let the curds set for 5 minutes.

6. Carefully pour off the whey to the level of the curds. Allow the curds to set for 30 minutes longer, stirring every 3 minutes to prevent matting. Maintain the temperature at 100°F (38°C).

7. Line a colander with cheesecloth and place it in a sink. Ladle the curds into the colander. Sprinkle the salt over the curds, mixing gently, and let drain.

8. Place the curds in a 2-pound mold lined with cheesecloth and press with a 3-pound weight for 15 minutes.

9. Remove the cheese from the mold and gently peel away the cheesecloth. Turn over the cheese, rewrap it, and press at 10 pounds of pressure for 12 hours.

10. Remove the cheese and place it on a clean surface at room temperature to air-dry. Turn twice daily until the surface is dry to the touch. Drying may take 1–3 days, depending on the temperature and humidity.

11. Wax the cheese (see page 50), if desired.

12. Age the cheese at 55°F (13°C) for 1–4 months, turning it at least once a week. (If you used raw milk, your cheese must be aged for at least 60 days.) The flavor will become sharper as the cheese ages.

Dry Jack

This aged Monterey Jack is a true American original, created in response to the needs of middle-class families in the 1950s who didn't have access to refrigerators. The result was an appealing aged cheese — crumbly, crowd-pleasing, and adaptable to many uses. The flavor of the cheese is helped along by a spicy rub of cocoa, pepper, coffee, and oil. This also makes for a spectacular presentation.

2 gallons milk

½ teaspoon calcium chloride diluted in ¼ cup cool, nonchlorinated water

1 packet direct-set mesophilic starter culture

½ teaspoon liquid rennet (or ½ rennet tablet) diluted in ¼ cup of cool, nonchlorinated water

2 tablespoons cheese salt

1 teaspoon espresso beans

2 tablespoons cocoa nibs

1½ teaspoons black peppercorns

3–4 tablespoons olive oil

Yield: 2 pounds

1. Slowly heat the milk to 90°F (32°C). Add the calcium chloride solution and stir well to combine. Sprinkle the starter over the surface of the milk, wait 2 minutes for the powder to rehydrate, then stir well. Cover the pot and let the milk ripen at 90°F for 45 minutes.

2. Add the diluted rennet and gently stir with an up-and-down motion for 30 seconds. Cover and let set undisturbed at 90°F (32°C) for 40 minutes, or until the curd gives a clean break.

3. Cut the curds vertically into 2-inch square columns. Let the whey rise from the cuts for 5 minutes. After 5 minutes, cut the curds into ⅜- to ½-inch chunks using a large whisk.

4. Over the next 45 minutes, stir gently as you slowly heat the curds to 102°F (39°C). The final result should be a curd that is firm throughout and moderately resistant when pressed between two fingers. Remove the pot from the heat and let the curds settle in the whey. Using a colander in the pot to hold down the curds, scoop out the whey until the liquid level is just at the level of the curds.

5. Wash the curds with 60°F (16°C) water, slowly adding just enough water (3 or 4 quarts) to cool the curds to 86°F (30°C), then continue to stir for 15 minutes. Remove the whey down to the curd level again.

6. Transfer the curd to a colander lined with butter muslin. Let the curd drain, stirring occasionally to help release the whey, for 10–20 minutes.

7. Add 1 tablespoon of the salt to the curd, stirring well to distribute. Let the curd rest and drain for 10 minutes, then add the remaining salt and stir again.

8. Gather the four corners of the butter muslin and pull them together to enclose the curd mass. Gently form the curd into a round ball by pulling the loose ends of the cloth up through your grip. As the curd begins to consolidate, apply pressure to the ball by rolling back and forth on a smooth surface to help mold it. It will continue to drain whey over this time. Tie off the cloth as tight and close to the ball as possible.

9. Place the ball knot side up on a cutting board. Top with a second board and a 25-pound weight. You will need to be creative in order to keep the weight and board level atop the round ball, so rig up a system that works. Let the cheese drain for 2 hours, then increase the weight to 50 pounds. Let it drain under the increased weight for 8–12 hours.

A cheese may disappoint. It may be dull, it may be naïve, it may be over-sophisticated. Yet it remains cheese, milk's leap toward immortality.

CLIFTON FADIMAN

Recipe continues on next page

10. The cheese will now be fully formed, with the curds entirely consolidated. When you remove the butter muslin, the top of the cheese will show the pattern of the cloth and a small button where the cloth was tied, but the bottom and the sides will be smooth. Let the cheese dry out at room temperature until the surface is dry to the touch, 5–10 days. Avoid too dry a space to prevent any cracking or curd separation. Turn the cheese daily, and brush off any mold that might appear.

11. One day before you anticipate the cheese's readiness, make the rub. Combine the espresso beans, cocoa nibs, and peppercorns in a spice grinder. Grind to a fine powder, then add enough oil to make a paste. Let the powder absorb the oil for 1 day. Transfer the cheese to a cooler aging space, at 52–56°F (11–13°C) and 80–85% relative humidity.

12. Rub the oil mixture into all the cracks and crevices of the top and sides of the cheese. Let the rub dry for a day, then flip the cheese and rub the other side with the mixture. Reapply the mixture 2 more times, waiting 2–3 days between applications.

13. Continue to age the cheese for 6–9 months.

AS A PRACTICING DAIRY-FARM VETERINARIAN for over two decades, Terry Homan witnessed the staggering decline in the number of family farms in western Wisconsin, where he himself had grown up on a dairy farm. Hence, in April of 2008, Homan launched Red Barn Family Farms to support local family farms (emphasis on family) by providing an additional revenue stream for their milk. It was also about honoring the animals. "This operation allows me to fulfill my professional oath — to protect the health and welfare of animals and alleviate their suffering, and to support conservation and public awareness — in the most wonderful way."

All Red Barn Family Farms must be certified by the American Humane Association and meet Terry's own rigorous standards of care (what he calls excellence in animal husbandry). "The better they take care of their cows, the cleaner the milk is, and the more they can earn for it. This is different from the commodity industry, which puts a premium on efficiency."

This being the Cheese State, Terry, who had zero cheese-making experience, tapped into the long-established infrastructure. Indeed, his very first phone call, to Wayne Hintz at Springside Cheese, resulted in the first product — a white cheddar — under the Red Barn label. "I'll be forever grateful because he answered the phone when I called, and we were a long shot. Hintz was used to producing 15,000 pounds of cheese and agreed to make a 5,000-pound vat for us." Hintz continues to produce RBFF's bandaged and block cheddars, which have garnered numerous awards over the years. Terry has also worked with the Center for Dairy Research at the University of Wisconsin as well as other notable cheese makers to develop a variety of cheeses.

Terry says one of the most rewarding aspects of RBFF has been the sense of appreciation the farmers feel when they meet people on the regular farm tours and how they now feel part of the future of farming. "Our dream is that someday the next layer of farms would seek to adjust the way they do things a bit, improve their animal care a bit, and then we would be having a very positive impact on the industry." For the time being, he is slowly growing his farm base, adding a new farm when the demand for the cheese dictates. "Long live the family farm!"

A CHEESE MAKER'S STORY

Terry Homan
RED BARN FAMILY FARMS
Western Wisconsin

*Amy, Steven, and Neal Holewinski
The Holewinski Family Farm
Pulaski, Wisconsin*

Baby Swiss

Switzerland produces a variety of Alpine cheeses, none of which are called Swiss cheese. Some, however, do have large holes — most notably Emmental, a large wheel with a mild taste. During the late nineteenth and early twentieth centuries, many Swiss cheese makers settled in the dairy belt of Wisconsin, and it was from this cultural collaboration that our Swiss cheese was born. This American version is quite a bit smaller than an Emmental, and so they called the nutty and creamy cheese "baby Swiss."

4 gallons milk

1 teaspoon calcium chloride diluted in ¼ cup cool, nonchlorinated water

1 packet C101 direct-set mesophilic starter culture or ⅜ teaspoon MM100 mesophilic starter culture

⅛ teaspoon *Propionic shermanii* powder

⅔ teaspoon liquid rennet (or ⅔ rennet tablet) diluted in ¼ cup cool, nonchlorinated water

1 gallon Saturated Brine (page 46)

Cheese salt

Light Brine (page 46)

Cheese wax (optional)

Note: To make this cheese with raw milk, decrease the MM100 starter culture to ¼ teaspoon. Increase the starter and rennet temperature to 86°F (30°C).

Yield: 4 pounds

1. Slowly heat the milk to 84°F (29°C). Add the calcium chloride solution and stir well to combine. Sprinkle the starter and *P. shermanii* over the surface of the milk and wait 2 minutes for the powders to rehydrate, then stir well to combine. Cover the pot and let the milk ripen at 84°F for 50 minutes.

2. Add the diluted rennet and slowly stir with an up-and-down motion for 30 seconds. Cover and let set undisturbed at 84°F for 40–45 minutes, or until the curd gives a clean break.

3. Cut the curd into ⅜-inch cubes (a wide whisk works well). Allow the curds to rest for 5 minutes, then stir gently for 5 minutes. After 5 minutes, let the curds settle to the bottom of the pot for 5 minutes.

4. Remove one-third of the whey. Stirring gently, slowly add enough 130°F (54°C) water to heat the curds to 95°F (35°C) over the next 5 minutes. Stir the curds for 5 minutes.

5. Slowly stir in enough more hot water to heat the curds to 102°F (39°C) over the next 5–10 minutes. At this point you have added back roughly the same quantity of water as the whey you removed.

6. Slowly stir the curds until they are dry and entirely cooked through, 30–40 minutes. Break a curd to test it. It should be firm throughout, with a moderate resistance when pressed between your fingers. Let the curds settle and stick together. Gather them on one side of the pot as they consolidate into one mass.

7. Using a colander in the pot to hold down the curds, scoop out the whey until the liquid level is just at the level of the curds. Place a plate on top of the curds, one large enough to cover the curds but small enough to fit into your pot. Top the plate with a 2½-pound weight, and let the curds rest for 20 minutes, or until they are well consolidated.

8. Remove the remaining whey and roll the curd mass into a piece of butter muslin. Transfer the bundle into a large hard-cheese mold and spread the excess butter muslin over the sides of the mold. Place a small piece of butter muslin between the curds and the follower and place the follower into the mold. While placing some pressure on the follower, pull up on the cloth in the mold to remove any folds; repeat this step when flipping the cheese during pressing.

Recipe continues on next page

9. Press the cheese at 10 pounds of pressure for 1 hour. Unwrap, flip, rewrap, and press at 10 pounds for another hour. Repeat every hour for a total press time of 5 hours. Keep the cheese at 75–80°F (24–27°C) throughout this time. If at any point the surface of the cheese is textured or bumpy, increase the weight to 25 pounds.

10. Transfer the cheese to a cooler space (52–56°F/11–13°C) to rest for 8–10 hours. Pay attention to the acid level of the cheese — it should not go below a final pH of 5.2–5.3, as excessive acid could impede the development of the gas-forming bacteria that will create the holes in your cheese. The final cheese should have a tight rind with no openings.

11. Soak the cheese in the saturated brine for 10–12 hours. Sprinkle the exposed surface with salt, flip the cheese halfway through the soak time, and sprinkle the other side with salt.

12. Remove the cheese from the brine and pat dry with sanitized cheesecloth.

13. Age the cheese in a cool place (50–55°F/10–13°C) at 75–80% relative humidity for 2–4 weeks. Turn the cheese daily and wipe it with light brine if mold starts to grow.

14. Transfer the cheese to an aging space at 65–70°F (18–21°C) and 80% relative humidity for 3–4 weeks of hole development. Turn the cheese daily to maintain even moisture and good hole distribution.

15. At this point, you can either wax the cheese (see page 50) or dry-brush it every few days to create a natural rind. Transfer the cheese to a cooler room (45–50°F/7–10°C) with 85% relative humidity for an additional month, or longer for a deeper and more developed flavor.

AS DEBRA HAHN SEES IT, "I feel like I have to be part of this book because my husband and I took a cheese-making class from Ricki about 18 years ago as part of a fun weekend getaway, and that's when I caught the bug!" The couple was gainfully employed at the time — she as a chemical engineer, he as a geologist — and neither had made cheese before. Nevertheless, on the ride back home he wondered, "Do you think one could actually make money doing that?" to which Deb responded, "No, never!"

Maybe not, but she definitely felt the pull to pursue it as a hobby. About five years later, when Debra was laid off from her job, she made cheese her livelihood, gaining as much knowledge as she could by attending classes and reading books. "Then you try all you learned with your own milk and in your own facility, and it's a whole other thing. My original cheese cave was only about 8 by 10 feet. Now I tell people to make their cave three times bigger than you think you need."

It took years of trial and error before she developed her signature cheese, The City of Ships. "It was the only one that I didn't have to struggle so much with. The first time I tried it I was like 'Oh, wow, I hit the nail on the head with this one.' Plus, I like aged cheeses the most, so that's kind of my niche." The aging part, too, took years to perfect, even after she built a larger cave in her cellar.

Like their names — others include Ragged Island and Eleanor Buttercup — the cheeses bear the stamp of their unique terroir. Her creamery-slash-home is perched on a peninsula with coastal fog, salty air, moderate climate, natural humidity, and other ideal cheese-making conditions — all qualities you can taste.

Debra has always used only raw milk for all her aged cheeses, something that was an oddity when she got started in 2000. "But the farmers love it when they see me coming every week, especially now that I'm buying 200 gallons of milk a week." Indeed, since the very beginning she's been buying milk from one family farm (Bisson), with four generations living on the land. "I'm really, really lucky."

A CHEESE MAKER'S STORY

Debra Hahn
HAHN'S END
Phippsburg, ME

BLUE CHEESES

Blue cheeses have been made since the time of the Romans. The term *Roquefort* first appeared in 1070 and referred to a blue-veined cheese made from ewe's milk. Blue cheeses are usually not pressed; they take their shape as the curds sink under their own weight in the mold. Salt is used for curing and to inhibit undesirable molds.

Blue cheeses are not for beginners. In many ways, a good homemade blue cheese is like a child: easy and fun to produce but difficult to bring to proper maturity. They must be aged in very damp, cool places for several months. If the proper environment is not maintained, they can dry out and develop undesirable surface molds. The first three recipes — blue, Stilton, and Gorgonzola — are presented in order of difficulty.

BLUE, *page 226*

BLEU D'AUVERGNE,
page 236

STILTON,
page 228

GORGONZOLA
DOLCE, *page 230*

225

BLUE CHEESE

*For the best flavor, use whole milk to make this cheese. Cow's milk makes a
light yellow cheese; goat's milk, a white one. The flavor comes from the mold
growing on the surface and inside the cheese.*

2 gallons milk

½ teaspoon calcium
chloride diluted
in ¼ cup cool,
nonchlorinated water

1 packet direct-set
mesophilic starter
culture

⅛ teaspoon *Penicillium
roqueforti* (blue mold)

½ teaspoon liquid
rennet (or ½ rennet
tablet) diluted
in ¼ cup cool,
nonchlorinated water

Cheese salt

Yield: 2 pounds

1. Heat the milk to 90°F (32°C) (86°F/30°C for goat's milk). Add the calcium chloride solution and stir well to combine. Sprinkle the starter and the *P. roqueforti* over the surface of the milk, wait 2 minutes for the powders to rehydrate, then stir well. Cover and let ripen at 90°F (86°F for goat's milk) for 60 minutes.

2. Add the diluted rennet and stir gently with an up-and-down motion for 30 seconds. Cover and let set at 90°F (32°C) (86°F/30°C for goat's milk) for 45 minutes, or until the curd gives a clean break.

3. Cut the curd into ½-inch cubes. Let set for 5 minutes at 90°F (32°C). If using goat's milk, let set for 10 minutes at 86°F (30°C).

4. Over the next 60 minutes, gently stir the curds every 3 minutes to keep them from matting.

5. Let the curds set undisturbed for 5 minutes.

6. Pour off the whey. Put the curds in a colander and let drain for 5 minutes. Put the curds back into the pot and gently mix them by hand so they are not matted.

7. Add 1½ tablespoons of the salt and mix well. Let set for 5 minutes.

8. Place a mat on a cheese board and the mold on the mat (the start of a mold sandwich; see page 44). Ladle the curds gently into the mold. Cover with a cheese mat and a cheese board (the finished mold sandwich). Let the curds drain for 12 hours. Turn over the mold every 15 minutes for the first 2 hours, then once an hour for the next 2 hours.

9. Remove the cheese from the mold; sprinkle the cheese with ½ tablespoon salt, distributing it evenly over all surfaces. Shake off excess salt. Let sit on a cheese mat or board at 60°F (16°C) and 85% relative humidity for 3–5 days. Turn it every day for 3 days, sprinkling the surface with salt and shaking off the excess each time.

10. Using a sterilized ice pick, $\frac{1}{16}$-inch stainless-steel knitting needle, or stainless-steel skewer, poke holes every inch from the top to the bottom of the cheese. Age at 50°F (10°C) and 95% relative humidity. It's best to store the cheese on edge, giving it a half turn daily. Mold will appear within 10 days.

11. After 20–30 days, mold will begin to appear on the surface. Scrape off excess mold every 20–30 days for 3 months.

12. After 3 months, wrap the cheese in foil and age at 42°F (6°C) and age another 60 days, turning the cheese weekly. Ideally, it is ready to eat after 6 months. For a milder cheese, sample after 3 months.

STILTON—STYLE CHEESE

This English cheese was first mentioned in the early part of the eighteenth century. It is made from milk enriched with cream and combines the subtle flavor of cheddar with the sharp taste of blue cheese.

2 gallons milk

2 cups light cream or half-and-half

½ teaspoon calcium chloride diluted in ¼ cup cool, nonchlorinated water

1 packet direct-set mesophilic starter culture

⅛ teaspoon *Penicillium roqueforti* (blue mold)

½ teaspoon liquid rennet (or ½ rennet tablet) diluted in ¼ cup cool, nonchlorinated water

2 tablespoons cheese salt

Yield: 2 pounds

1. Thoroughly combine the milk and cream, and heat the mixture to 86°F (30°C). Add the calcium chloride solution and stir well to combine. Sprinkle the starter and *P. roqueforti* over the surface of the milk, wait 2 minutes for the powders to rehydrate, then stir well. Cover and allow the mixture to ripen for 30 minutes.

2. Add the diluted rennet and stir gently with an up-and-down motion for 30 seconds. Top-stir for another 30 seconds if you see any butterfat rising. Cover and let set at 86°F (30°C) for 90 minutes.

3. Using a slotted spoon, transfer the curds into a cheesecloth-lined colander resting in a bowl. When the curds are in the colander, they should be surrounded by whey in the bowl. Let the curds sit in the whey for 90 minutes.

4. Tie the corners of the cheesecloth into a knot and hang the bag to drain until it stops dripping, about 30 minutes.

5. When the bag has stopped dripping, place it on a cheese board on a tray to catch the whey while pressing. Place a board on top of the bag and place 8 pounds of weight on the board. Press the cheese at 68–70°F (20–21°C) for 8–12 hours.

6. Remove the curds from the bag. Break into 1-inch pieces and put them in a bowl.

7. Add the salt and gently mix.

8. Place a mat on a cheese board and a 2- to 4-pound hard-cheese mold on the mat (the start of a mold sandwich; see page 44). Ladle the curds gently into the mold. Place a mat on top of the mold and a cheese board on top of the mat (the finished mold sandwich).

9. Carefully flip over the mold every 15 minutes for 2 hours. Let sit at 70°F (21°C) for 8–12 hours.

10. Over the next 4 days, flip the cheese several times a day, maintaining it at 70°F (21°C).

11. Using a sterilized ice pick, ¹⁄₁₆-inch stainless-steel knitting needle, or stainless-steel skewer, poke holes every inch from the top to the bottom of the cheese.

12. Age the cheese for 4 months on a cheese mat at 50–55°F (10–13°C) with a relative humidity of 90%. Turn the cheese twice weekly and scrape off any mold or slime once a week. Repierce any holes that become clogged.

GORGONZOLA DOLCE

Gorgonzola Dolce originated in the northern Italian regions of Lombardy and Piedmont, and yes, Gorgonzola is a real place. The cheese is originally from the mountains and is now made in the lowlands along the Po River Valley. A moister version of Gorgonzola made today accounts for about 80 percent of the market for all Gorgonzola cheese. This deliciously creamy cheese is sweet and spreadable; it ages rather quickly compared to its blue cousins.

1⁄16 teaspoon *Penicillium roqueforti* (blue mold) diluted in ½ cup of cool milk

2 gallons milk

½ teaspoon calcium chloride diluted in ¼ cup cool, nonchlorinated water

1 packet direct-set buttermilk starter culture

⅓ cup prepared Bulgarian yogurt

½ teaspoon liquid rennet (or ½ rennet tablet) diluted in ¼ cup cool, nonchlorinated water

Cheese salt

Yield: 2 pounds

1. Make the *P. roqueforti* solution and let it sit for 30 minutes while heating the milk to 90°F (32°C). Add the calcium chloride solution to the milk and stir well to combine. Sprinkle the starter over the surface of the milk, wait 2 minutes for the powder to rehydrate, then stir well. Add the Bulgarian yogurt and *P. roqueforti* solution and stir well. Cover and allow the milk to ripen for 60 minutes.

2. Add the diluted rennet and stir gently with an up-and-down motion for 30 seconds. Top-stir for another 30 seconds if you see any butterfat rising. Cover and let set at 90°F (32°C) for 30 minutes, or until the curd gives a clean break.

3. Keeping the curds at 90°F (32°C), cut the curds into 1-inch cubes. Stir every 3 minutes for 20 minutes to prevent the curds from matting. Using a colander in the pot to hold down the curds, remove 2 quarts of whey. Stir every 3 minutes for another 20 minutes to prevent the curds from matting.

4. Gently ladle the curds into a cheesecloth-lined colander. Let drain for several minutes, gently lifting the corners of the cheesecloth to facilitate draining. The goal is to keep the curds separate, rather than consolidating them, to encourage mold growth.

5. Start a mold sandwich (page 44) with an unlined 2-pound hard-cheese mold. Ladle the curds gently into the mold. After 5 minutes, flip the sandwich. Drain at 80–90°F (27–32°C), flipping every 20 minutes for the first hour then every hour for the next 4 hours. There may be some openings in the cheese at this point, which is expected.

6. Remove the cheese from the mold and gently rub 1 teaspoon of salt onto the top and sides of the cheese, then return it to the mold. After 1 day rub the other side of the cheese with 1 teaspoon salt and return it to the mold with the freshly salted side on top. Repeat this process for 2 days.

7. Age at 52–54°F (11–12°C) and 93–95% relative humidity for 7–10 days. Using a sterilized ice pick, $\frac{1}{16}$-inch stainless-steel knitting needle, or stainless-steel skewer, poke holes every inch from the top to the bottom of the cheese. Age for another 90 days.

SHROPSHIRE

SHROPSHIRE

This mild and crumbly cheese is one of the lesser-known blues, but it's also one of the most delicious. Shropshire gets its bright orange color from natural annatto coloring, and although it looks like Stilton, it's much creamier and less nutty-tasting. Start this cheese early in the day, or you'll be working late into the night.

¾–1½ teaspoons annatto cheese coloring diluted in ¼ cup cool, nonchlorinated water

2 gallons milk

¹⁄₁₆ teaspoon *Penicillium roqueforti* (blue mold)

½ cup cooled, sterilized water

½ teaspoon calcium chloride diluted in ¼ cup cool, nonchlorinated water

½ packet C101 direct-set mesophilic starter culture or ⅛ teaspoon MA11 mesophilic starter culture

¼ teaspoon liquid rennet (or ¼ rennet tablet) diluted in ¼ cup cool, nonchlorinated water

2½ tablespoons cheese salt

Yield: 2 pounds

1. Stir the annatto solution into the milk. Use the lesser amount for a lighter-colored cheese and more for a brighter orange. You will not see the final color until the cheese is drained and finished.

2. Start to slowly heat the milk to 88°F (31°C). While the milk heats, rehydrate the *P. roqueforti* in the sterilized water and let it sit for 30 minutes.

3. Add the calcium chloride solution to the milk and stir well to combine.

Recipe continues on next page

4. When the milk is at 88°F (31°C), sprinkle the starter over the surface of the milk, wait 2 minutes for the powder to rehydrate, then stir well. Add the *P. roqueforti* solution and stir. Cover the pot and let the milk ripen at 88°F for 30 minutes.

5. Add the diluted rennet and stir gently with an up-and-down motion for 30 seconds. Cover and let set undisturbed at 88°F (31°C) for 90 minutes, or until the curd gives a clean break.

6. Cut the curd vertically into 1-inch-square columns. There are no horizontal cuts for this cheese. Let the curd sit for 3 hours without stirring, continuing to keep the pot at 88°F (31°C). The curd will condense and settle in the whey over this time.

7. Line a colander with butter muslin and set the colander within a deep pan or pot. Remove the whey to the level of the curd. Gently transfer the curd and the remaining whey to the colander. The curd will be quite delicate; do your best to keep the pieces whole as you transfer them. The whey will eventually submerge the curd entirely, keeping it warm and supplying it with lactose. Let it drain for 1 hour, then remove the remaining whey.

8. Gather the corners of the cloth together, tying them in a knot to cinch the fabric tightly around the ball of curd. Turn the ball over and replace it in the colander, knot side down. Place the colander in the pot and let sit for another hour.

9. Remove the colander from the pot. Untie the cloth, flip the curd, retie, and return to colander. Transfer the colander to a draining board and let the curd drain for another hour.

10. Unwrap the curd from the cloth and cut it into 2- to 3-inch cubes. Transfer the cubes to a cheese mat on a draining board. Keep the curd at 80°F (27°C) and let it continue to drain for 1 hour, turning the pieces every 15 minutes. Let the curd rest for 12 hours, cooling to 72°F (22°C) over this time.

11. Break the curd pieces into quarters. The torn curd should have a flaky texture. If checking the pH, it should be 4.6–4.7. Add one-third of the salt to the curd and mix until it dissolves. Add another third of the salt and mix until dissolved, then repeat with the remaining salt.

12. Make a mold sandwich with a 4-pound cheese mold (page 44). Pack the curds into the mold. Let the curd drain at 60–65°F (16–18°C) for 5 days, flipping daily. During this time, the curds will consolidate to one-half to one-third of their initial height. If the curd is too dry and not consolidating well after the second day, add a follower and 4–8 pounds of weight.

13. Unmold the cheese. It will be firm, but will have large cracks in the surface. Dip the dull edge of a knife in warm water, and scrape it back and forth over the surface to fill in the cracks. Cut a dry piece of butter muslin long and wide enough to wrap around the cheese. Cut two more squares of cloth for the top and bottom of the cylinder, and wrap one cheese. The cloth will continue to wick moisture away from the cheese as it dries. Let the cheese dry in a 60°F (16°C) room with moderate moisture, turning it daily until the bandage dries out and blue mold appears on the surface. This will take 2–4 days.

14. Remove the bandage and age at 52–54°F (11–12°C) and 85% relative humidity. After 5 weeks, use a sterilized ice pick, $\frac{1}{16}$-inch stainless-steel knitting needle, or stainless-steel skewer to poke holes every inch from the top to the bottom of the cheese, going about two-thirds of the way into the body of the cheese. Age the cheese until the surface becomes dark and wrinkled, 3–6 months.

Bleu d'Auvergne

This creamy and crowd-pleasing blue originates in the volcanic mountains in the middle of southern France, in an area known as the Massif Central. The volcanic soil gives the region's milk a deep richness, a quality that originally gave Bleu d'Auvergne its character. Although similar to Roquefort, this blue is milder, creamier, less salty, and more approachable. It is an advanced cheese, so be sure to make a few other cheeses before attempting it.

2 gallons milk

¹⁄₁₆ teaspoon *Penicillium roqueforti* (blue mold)

2 tablespoons cooled, sterilized water

½ teaspoon calcium chloride diluted in ¼ cup cool, nonchlorinated water

¹⁄₁₆ teaspoon MM100 mesophilic starter culture

¹⁄₃₂ teaspoon TA61 thermophilic starter culture

⅛ teaspoon liquid rennet (or ⅛ rennet tablet) diluted in ¼ cup cool, nonchlorinated water

Cheese salt

Yield: 2–3 pounds

1. Slowly heat the milk to 90°F (32°C). Rehydrate the *P. roqueforti* in the sterilized water. Add the calcium chloride solution to the milk and stir well to combine. Sprinkle the starters over the surface of the milk, wait 2 minutes for the powders to rehydrate, and stir well. Add the *P. roqueforti* solution and stir well. Cover the pot and let the milk ripen for 60 minutes at 90°F.

2. Add the diluted rennet and stir gently with an up-and-down motion for 30 seconds. Cover and let set undisturbed for 90 minutes, or until the curd gives a clean break.

3. Meanwhile, set a large draining tray at an angle and line it with butter muslin.

4. Cut the curd into ½- to ¾-inch cubes and let rest for 5 minutes. Stir intermittently for 20 minutes, stirring just enough to firm up the curds and keep them from sticking together. Scoop out about one-fifth of the whey, then stir gently for 10–15 minutes. Scoop out an additional one-fifth of the whey, and again stir for 10–15 minutes. Check the curds at this point. They should be plump and springy, with a light skin on the surface. They shouldn't be so wet that they stick together, and they shouldn't be so dry that they're firm all the way through.

5. When the curds are ready, give them a quick and gentle stir and transfer them into as thin of a layer as possible on the cloth-lined tray. Let them drain for about 15 minutes, occasionally stirring to keep the curds from sticking together. It's essential to keep the curds separate throughout the draining process, as the curd structure will help to create the spaces for the blue mold to develop during the aging process.

6. Make a mold sandwich (page 44) with a large hard-cheese mold. Ladle the curds into the mold and flip after 15 minutes. Add a follower and a 4–5 pound weight, and let sit at 72°F (22°C) for 3–4 hours, flipping every 30 minutes. Transfer the cheese to a cooler space (52–54°F/11–12°C) to rest for 12 hours.

7. Remove the cheese from the mold and spread 1 tablespoon of salt over the top of the cheese all the way to the edges, rubbing the excess into the sides of the cheese. Return the cheese to the mold and let it sit for 24 hours as it creates its own brine and reabsorbs it. Flip the cheese, and repeat the process with an additional tablespoon of salt, then let sit for 24 hours. Flip it a third time and rub a tablespoon of salt over the entire surface of the cheese, for a total salting time of 3 days. The surface will become drier and harder over this time.

8. Transfer the cheese to a large plastic container with a cover to maintain a high moisture level through the next step of the process. (A plastic cake carrier is ideal.) Hold the cheese at 46–54°F (8–12°C) with 90–95% relative humidity for 1 week, turning it every day. The surface of the cheese will soften over this time as the salt penetrates it and the interior moisture migrates to the surface.

Recipe continues on next page

9. Use a sterilized ice pick, $\frac{1}{16}$-inch stainless-steel knitting needle, or stainless-steel skewer to poke holes every inch from the top to the bottom of the cheese.

10. Let the cheese age in the same cool and humid space for 60–75 days, flipping daily. After 3 weeks, you will begin to see signs of blue mold. If the humidity level is too high, a rose or orange surface will develop, and if the humidity levels is too low, a dry white surface mold will develop.

BLEU D'AUVERGNE

CAMBLU

This hybrid cheese brings together the white rind and soft cream of a Camembert with a mild blue cheese. When it's perfectly ripe, the blue melts into the Camembert, similar to the French Bleu de Bresse.

2 gallons milk

12 ounces heavy cream

½ teaspoon calcium chloride diluted in ¼ cup cool, nonchlorinated water

1 packet C21 buttermilk starter culture or ¼ teaspoon MM100 mesophilic starter culture or ¼ teaspoon Flora Danica* starter culture

¼ teaspoon liquid rennet (or ¼ rennet tablet) diluted in ¼ cup cool, nonchlorinated water

¼ teaspoon *Penicillium roqueforti* (blue mold)

Cheese salt

WHITE MOLD SOLUTION

¼ teaspoon cheese salt

⅛ teaspoon *Penicillium candidum*

1/32 teaspoon *Geotrichum candidum*

1 cup cool, sterilized water

*Flora Danica may be used as either a direct-set or a reculturable starter (see page 15).

Yield: 3 pounds

1. Combine the milk and cream and slowly heat the mixture to 90°F (32°C). Add the calcium chloride solution and stir well to combine. Sprinkle the buttermilk culture over the surface of the milk, wait 2 minutes for the powder to rehydrate, then stir well. Cover the pot and let the milk ripen at 90°F for 30 minutes.

Recipe continues on next page

2. Add the diluted rennet and gently stir with an up-and-down motion for 30 seconds. Cover and let set undisturbed at 90°F (32°C) for 90 minutes, or until the curd gives a clean break.

3. Cut the curd into ¾-inch cubes. Stir very gently and slowly at 90°F (32°C) for 20–30 minutes. The desired result is a curd with a dry exterior, an outer skin, and a high moisture content.

4. Let the curds settle while you make three mold sandwiches using Camembert molds (page 44). With a ladle, remove the whey to the level of the curds. Use a slotted spoon to begin the transfer of curd to the molds, taking care to shake off excess whey as you go. Build a level of curd about ¾ inch high in each mold, making sure the base is entirely covered.

5. Sprinkle a pinch (1/64 teaspoon) of the *P. roqueforti* across the surface of each cheese, avoiding the outer half inch.

6. Transfer more curds to the cheese forms, building up an additional ¾-inch layer of curds. Repeat the sprinkling process, alternating layers of curd and *P. roqueforti*, finishing with a final layer of curd to encapsulate the blue mold inside the cheese.

7. Keep the curds warm (68–72°F/20–22°C) for the next 10–12 hours, gently flipping the molds every 2–3 hours. The curds will shrink to about half of the original height. When fully drained, the sides of the cheese will have a slightly open texture.

8. Remove each cheese from its mold. Spread ½ teaspoon of cheese salt evenly over the top surface of each cheese. Rub the salt over the edge of each block and spread the excess over the sides. Return each cheese to its mold salt side up. Let them sit for 4–6 hours, then flip and salt each of the other sides with an additional ½ teaspoon of salt. Return the cheeses to the molds, and let them sit for an additional 4–6 hours.

9. Transfer the cheeses from the molds to a dry surface in a space at a temperature of 58–65°F (14–18°C)and 60–70% relative humidity. Let the cheeses rest and dry, turning a few times a day, until the surface dries, 1–2 days.

10. Once the cheese is dry, the aging process begins. Transfer the cheeses to an enclosed space, such as a plastic container with a lid, at a temperature of 52–56°F (11–13°C) and 92–95% relative humidity. Let the cheese rest, turning twice daily, for 2 days.

11. Use a sterilized ice pick, $\frac{1}{16}$-inch stainless-steel knitting needle, or stainless-steel skewer to poke holes every inch from the top to the bottom of each cheese, inserting it at least halfway into the cheese. Let the cheeses rest for 1–2 days.

12. To make the white mold solution, combine the ¼ teaspoon salt, the *P. candidum*, and the *G. candidum* with ½ cup of the cool sterilized water and let the mold powders rehydrate for 8 hours. Using a very fine spray bottle, mist the surface of each cheese with the white mold solution, taking care not to get the cheese too wet. Allow the surface to dry slightly, then return the cheeses to the aging space.

13. Let the cheeses ripen, turning them once or twice daily. In 7–10 days, the cheeses should be fully covered with white mold. Now age the cheeses at a temperature of 42–46°F (6–8°C) with 92–95% relative humidity. If the white mold begins to grow over the aeration holes, pierce the surface again to enable the blue growth. Continue to age the cheese, monitoring and flipping daily, for 30–45 days. The cheeses are ready when they're soft throughout, with fully developed mold growth.

Jill Giacomini

**POINT REYES
FARMSTEAD CHEESE
COMPANY**

Point Reyes, California

Jill, Lynn, and Diana Giacomini

HAPPY HOLSTEINS, INDEED: Point Reyes Creamery is located on 720 acres that hug the Pacific Coast Highway, about an hour's drive north of San Francisco. Here in Tomales Bay, cheese making is a long-standing tradition. "My dad was raised on a dairy farm nearby," says Jill, who owns the farmstead operation along with her sisters Karen (now retired), Diana, and Lynn. "After he graduated from UC Davis, he and my mom bought this farm in 1959 and it, too, was a working dairy farm, though we girls were encouraged to develop our own interests."

Fast-forward to the 1990s, when the sisters are all in one form of business or another and felt free to come back on their own initiative — and to make a reality their dad's dream of creating a branded product that he could sell directly to the consumer. "For the first 40 years he watched the milk truck pull away every morning and saw his hard-earned milk processed into different products, all sold under another name," Jill says, explaining how running such a massive farm prevented him from pursuing that dream himself. "We knew the only way to do that would be to diversify into some other operating business, so okay, let's make cheese, which is what dad has always wanted to do, and we'll do it together."

Jill cautions new cheese makers to resist the urge to start off producing too many cheeses when you are also trying to build your business, and emphasizes the importance of hiring the right people and concentrating on food safety. "A lot of people told us we would never be able to build a brand on just one cheese, but we did," she says, adding that Original Blue is still their flagship, though it continued to be perfected over the next eight or so years before they hired their current cheese maker, Kuba Himmerling, to grow the product line.

Now they offer fresh mozzarella, Tomo (a farmer's cheese), aged Gouda, and other types of blue cheeses. "If our cheese-making team does their job right, all our cheeses taste like our milk, and our milk tastes like West Marin." For Jill and her clan, that's the sweet measure of success.

8

GOAT'S-MILK CHEESES

lthough nearly all of the cheeses in this book may be made with goat's milk, this chapter is dedicated to the goat lovers out there who are looking for a few extra "wheys" to use their own milk. You can also use cow's milk for any of the recipes in this chapter, but it won't produce the characteristic "goaty" flavor that comes from the fatty acids — capric, caproic, and caprylic — present in goat's milk. These cheeses combine the techniques used for hard, soft, and mold-ripened cheeses; to review those techniques, see chapter 3.

Goat's-milk curd is softer than that of cow's milk, so it must be treated very gently. After cutting the curd, you may have to let the cubes settle for 10 minutes so they firm up enough to begin the cooking process. Remember that because goat's milk doesn't contain carotene, it produces a white cheese. You may add cheese coloring to the milk if you prefer. It will not change the flavor or the texture of the cheese.

Note: Flora Danica starter is an excellent fresh goat-cheese starter for any of the soft-cheese recipes. It may be used either as a direct-set or as a reculturable starter (see page 15).

CHÈVRE,
page 249

243

Soft Goat Cheese

This delicious soft cheese is a great place to start making cheese. You may want to experiment with various sizes and shapes of soft-cheese molds. It makes a terrific spread for sandwiches, bagels, and crackers.

For a delicious variation, try adding dried herbs in between the layers of curd in the soft goat cheese recipe. Make sure they are dried and sterilized. Some great flavors are dill, caraway seeds, and freshly ground black pepper. You can also roll the final cheese in herbs to impart a flavorful outside with a delectable smooth finish inside.

1 gallon goat's milk

½ teaspoon calcium chloride diluted in ¼ cup cool, nonchlorinated water

1 packet buttermilk direct-set starter culture

1 drop liquid rennet diluted in 5 tablespoons cool, nonchlorinated water

1 teaspoon cheese salt (optional)

Yield: About 1 pound

1. Heat the milk to 86°F (30°C). Add the calcium chloride solution and stir well. Sprinkle the starter over the surface of the milk, wait 2 minutes for the powder to rehydrate, then stir well to combine.

2. Add the diluted rennet solution and stir gently with an up-and-down motion for 30 seconds.

3. Cover and allow the milk to set at 72°F (22°C) for 12–24 hours, or until you have a firm coagulation.

4. Place four to eight soft-cheese molds on a draining mat set on a wire rack over a basin to collect the whey. Gently ladle the curds into the molds round-robin style, being careful not to break up the curd. Fill the molds, wait 15 minutes for the curd to settle, and ladle in more curd until it is all used up. Let the cheese drain for 18–24 hours, turning once to help drainage. It will settle by one-third to one-half of its volume.

5. Unmold the cheese. If desired, lightly salt the surface immediately after unmolding. Eat immediately or wrap it in cheese wrap and store for up to 2 weeks in the refrigerator.

SAINTE-MAURE DE TOURAINE

Originating in the Loire region of France over one thousand years ago, this cheese is characterized by its log shape, mold growth, and often pungent flavor.

2 gallons goat's milk

½ teaspoon calcium chloride diluted in ¼ cup cool, nonchlorinated water

1 packet chèvre starter culture

¹⁄₁₆ teaspoon *Geotrichum candidum* (white mold)

4 pieces rye straw (optional)

Cheese salt

Chopped dried herbs or edible flower petals (optional)

Ash (optional)

Yield: About 1 pound

1. Heat the milk to 86°F (30°C). Add the calcium chloride solution and stir well to combine. Sprinkle the starter and the *G. candidum* over the surface of the milk, wait 2 minutes for the powders to rehydrate, then stir well.

2. Cover and allow the milk to set at 72°F (22°C) for 18–24 hours, or until there is a ⅛- to ¼-inch layer of whey on the surface of the curd.

3. Set up four Sainte-Maure cheese molds on a draining mat. Keep them upright by wrapping them together with a couple of large rubber bands. A fifth empty mold adds stability. (See illustration at right.) Set the mat on a wire rack in a basin or in the sink, to collect the whey as it drains. Very gently remove the whey from the surface of the curd. Carefully spoon or ladle thin slices of curd into the molds, keeping the curd as intact as possible. Fill the molds to the top, round-robin style, wait 15 minutes for the curds to settle, and repeat this process until all curd is in the molds.

Recipe continues on next page

4. Put the molds in a convenient place to drain at 72°F (22°C) for 18–24 hours. The curd will settle by at least one-half of its volume. Even when the whey stops dripping, the cheese is still quite fragile.

5. It is traditional to put a piece of rye straw through the center for ease of handling, and to provide strength and some aeration for the center of the cheese. If you can find some, sanitize it and insert a piece into each cheese before carefully removing it from the mold.

6. You have three options for finishing this cheese:
   - Roll each cheese in 1 teaspoon of cheese salt spread on a piece of wax paper.
   - After the salt has been absorbed, you can roll the cheeses in finely chopped dried herbs, edible flower petals, or whatever delicious combination you prefer.
   - Roll each cheese in a mixture of 1 teaspoon salt and 1 teaspoon ash. The ash will sweeten the surface of the cheese by neutralizing the acid it contains, and will help draw out moisture and keep the rind from becoming sticky during aging.

7. Let the cheeses dry at 72°F (21°C) for 2–3 days, then age them at 52–56°F (11–13°C) with 90% relative humidity. Turn them daily for 6–21 days. If you are making this recipe for the first time, try the cheeses at different stages to find out what they taste like at different aging levels.

"IT ALL STARTED IN HIGH SCHOOL when my dad, a peanut farmer, bought me two goats," explains Gail, who grew up on 125 acres in North Carolina. "I would milk the goats before school and there was so much milk we had to do something with it, so I made yogurt and ricotta and cottage cheese under the tutelage of my grandmothers."

Not that cheese making was her first career move. She spent years building her résumé as a chef at notable establishments in the South before meeting her husband and buying a small farm in Esmont, in central Virginia, where eventually she set up shop. (Caromont is a nod to Carolina and Esmont.) "I started off by milking 13 goats two at a time into a bucket, then carting the buckets in a Radio Flyer wagon to the cheese-making room — a plastic shed — where I would pour the milk into my JayBee vat pasteurizer, throw in a handful of culture, and hope for the best." She was selling 300 pounds of cheese a year to restaurants and cheese shops and local wineries. "That was my dream phase."

Four years later, bored with making fresh chèvre, she was ready to throw in the towel. "I wanted to replicate these little village cheeses from the Loire Valley that I had tasted as a chef at the Inn at Little Washington." So she attended some intensive courses, created production plans and budgets, and built a milking parlor. Now she makes 35,000 pounds of cheese each year in a variety of fresh and aged types, including some with cow's milk to help pay the bills during the winter.

Gail also finds gratification in using the farm for other than cheese making, holding planned "snuggle sessions" where visitors can mingle with the goats. When the first Farm Therapy event in 2016 brought 2,000 people from all across the country, she decided to invite specific agencies (including those dealing with child trafficking) to provide more focused therapy sessions. "It is so uplifting to see these folks go from being quiet and reserved to laughing out loud when a baby goat jumps up on their table." Gail sees the farm being used for outreach a lot more in the future, with the snuggle sessions providing the funding for those programs. "It's a way to blend my political life and my farm life, especially as a woman farmer."

A CHEESE MAKER'S STORY

Gail Hobbs Page
CAROMONT FARM
Esmont, Virginia

DRY COTTAGE CHEESE

Here is a delicious variation on cottage cheese made from goat's milk. It is a washed-curd cheese, therefore much of the lactose is removed.

1 gallon goat's milk

¼ teaspoon calcium chloride diluted in ¼ cup cool, nonchlorinated water

1 packet chèvre starter culture or 1 packet direct set buttermilk starter culture

3 drops liquid rennet diluted in ¼ cup cool, nonchlorinated water (only if using buttermilk starter culture)

Cheese salt (optional)

Herbs (optional)

Yield: 1 pound

1. Heat the milk to 72°F (23°C). Add the calcium chloride solution and stir well to combine. Sprinkle the starter over the surface of the milk, wait 2 minutes for the powder to rehydrate, then stir well.

2. If you are using the chèvre culture, omit this step. If you are using the buttermilk starter, add the diluted rennet and stir gently with an up-and-down motion for 30 seconds.

3. Cover and let set at 72°F (22°C) for 24 hours. Cut the curd into ½-inch cubes. Allow them to rest for 5 minutes.

4. Gradually heat the curds to 116°F (47°C), raising the temperature 5 degrees every 5 minutes for 40 minutes, stirring every 3 minutes to prevent the curds from matting.

5. Let the curds rest for 5 minutes at 116°F (47°C). When ready, the curd will be moderately resistant when pressed between your fingers.

6. Using a colander in the pot to hold down the curds, ladle off the whey, then transfer the curds into a butter muslin–lined colander, stirring gently every 3 minutes to prevent the curds from matting. Drain for 10 minutes.

7. Remove the butter muslin and submerge the colander of curd in a pot of cold, sterilized water to remove lactose from the curds. When the water gets milky, drain it off and replace it. When the water runs clear, drain and store the cheese in a covered container in the fridge for up to 10 days. If you wish to add salt or herbs, mix in to taste before eating.

NOTE: You can also press the curds in a basket mold and at 10 pounds for 20 minutes. Flip and press at 10–15 pounds for another 20 minutes.

CHÈVRE

*Chèvre (French for "goat") is easy to make. This fresh, creamy cheese is an
excellent spread, and you can add herbs and spices for variety. Or use it as a
substitute for cream cheese or ricotta in cooking. When made with whole milk,
it has the consistency of cream cheese. Chèvre made with skim milk has a lower
yield and a drier consistency.*

1 gallon goat's milk

¼ teaspoon calcium
chloride diluted
in ¼ cup cool,
nonchlorinated water

1 packet direct-set
chèvre starter culture

Herbs or spices
(optional)

Cheese salt (optional)

Yield: 1½ pounds

1. Heat the milk to 86°F (30°C). Add the calcium chloride solution and stir well to combine. Sprinkle the starter over the surface of the milk, wait 2 minutes for the powder to rehydrate, then stir well.

2. Cover and let set at 72–86°F (22–30°C) for 12 hours. (I like to make this cheese at night, so I can drain it when I get up in the morning.)

3. Line a colander with butter muslin. Gently ladle the curds into the colander. Tie the corners of the muslin into a knot and hang the bag over the sink to drain for 3–5 hours, or until the curds reach the desired consistency. The shorter the draining time, the creamier the cheese. A room temperature of at least 72°F (22°C) will encourage proper draining.

4. Add herbs, spices, or salt, if desired, and mix in. Store in a covered container in the refrigerator for up to 1 week, draining any excess whey that forms.

SOFT GOAT
CHEESE
WITH HERBS,
page 244

CHÈVRE, *page 249*

GOAT'S-MILK
CHEDDAR, *page 253*

SAINTE-MAURE
DE TOURAINE,
page 245

Laini Fondiller

LAZY LADY FARM

Westfield, Vermont

A FAVORITE AMONG CHEESEMONGERS, CHEFS, AND CUSTOMERS ALIKE, Lazy Lady Farm produces about 10,000 pounds of cheese a year — a far cry from the early days when Laini Fondiller was plying her "misshapen" cheeses at the local farmers' market. Like many American cheese makers, Laini caught the bug during a stint abroad. "I fell in love with the process as well as the world of cheese while working on dairy farms in Corsica, France. It was eye-opening to see something being made from milk."

Upon returning home to Vermont, she bought some land and a couple of goats and, being without electricity or money to invest in a cheese-making facility, began producing cheese in the kitchen, lining a closet with plastic for draining and curing. "I always wrote things down and fiddled with temperatures and cuttings and stirring versus nonstirring. Lots and lots and *lots* of trial and error. I had all that French cheese in mind but didn't know how to get there. There was no Internet, there were no experts to call up, so I just had to learn as I went."

Eventually she built a 10-by-14-foot room so she could pass inspection, created her own five-gallon pasteurizer, and built a geothermal cave. "We stumbled and bumbled our way in the beginning. I had to learn how to control the flora and fauna."

In 2002, thanks to the Vermont Community Loan Fund, she built a 20-by-24-foot room and bought a 50-gallon vat, thereby doubling her production of milk to about 300 gallons per week. She built a second cave in 2008, which is where she ages the bloomy and washed rinds, keeping the aged cheeses in the original cave.

Laini continues to produce cheese on a cyclical basis (and without electricity to power a dehumidifier or an air conditioner). Her regular roster includes mostly lactic-acid types of goat cheese, plus cheeses using cow's milk from neighboring farms during the off-season, so she can keep the cash flow going and hire employees year-round. "I work 98 hours a week without a day off. I can't afford to scale back because I would have to lay off someone, and I have a responsibility to those employees. But cheese still blows my mind: the world of opportunity milk provides to you and the twists and turns it throws at you. It will do stuff to you."

GOAT'S-MILK CHEDDAR

This stirred-curd variety of cheddar has a terrific sharp, peppery flavor. While it's ready for eating after just four weeks, the flavor will improve if aged longer.

2 gallons goat's milk

½ teaspoon calcium chloride diluted in ¼ cup cool, nonchlorinated water

1 packet direct-set mesophilic starter culture

½ teaspoon liquid rennet (or ½ rennet tablet) diluted in ¼ cup cool, nonchlorinated water

2 tablespoons plus 1 teaspoon cheese salt

Cheese wax (optional)

Yield: 2 pounds

TROUBLESHOOTING

If the curds are not setting firmly enough for you to cut easily, next time add ½ teaspoon calcium chloride diluted in ¼ cup water to the milk before adding the starter.

1. Heat the milk to 85°F (29°C). Add the calcium chloride solution and stir well to combine. Sprinkle the starter over the surface of the milk, wait 2 minutes for the powder to rehydrate, and stir well. Cover and allow the milk to ripen for 30 minutes.

2. Add the diluted rennet and stir gently with an up-and-down motion for 30 seconds. Cover the pot and let set for 1 hour at 85°F (29°C).

3. Cut the curd into ½-inch cubes. Allow the curds to sit undisturbed for 10 minutes.

4. Gradually heat the curds 2 degrees every 5 minutes to 98°F (37°C). Stir gently every 3 minutes to keep the curds from matting. Maintain the temperature at 98°F for 45 minutes, stirring gently every 3 minutes. Remove the whey. Add 2 tablespoons of the salt to the curds and mix.

5. Line a 2-pound cheese mold with cheesecloth. Ladle the curds into the mold. Press the cheese at 20 pounds of pressure for 15 minutes.

6. Remove the cheese from the mold; gently peel away the cheesecloth. Flip the cheese, rewrap it, put it back into the mold, and press at 30 pounds of pressure for 1 hour.

7. Unwrap, flip, rewrap, and press at 50 pounds of pressure for 12 hours. Remove from the press and gently peel away the cheesecloth. Rub salt on all surfaces. Place on a cheese board.

8. Turn the cheese daily for 2 days, rubbing salt on it once a day at room temperature. When the surface is dry, you may wax it for a moister cheese (see page 50) or age it naturally for a drier, more flavorful cheese.

9. Age the cheese at 50–55°F (10–13°C) for 4–12 weeks.

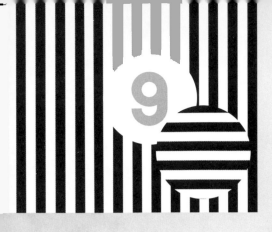

WHEY CHEESES

Prepare yourself: When you make hard cheese, there is usually a hefty amount of whey left over. Being that it's packed with albuminous proteins and minerals, it would be a shame to just throw it away, and it is too acidic to be composted. That's why so many farmstead cheese makers have their own pigs; I've heard that if you feed whey to pigs, you can reduce your grain bill by one-third.

Even if you don't have pigs, you can put all that whey to good use. Not surprisingly, cheese makers have devised methods to use whey to make more cheese. Although the temperatures used in hard-cheese making are not high enough to separate out the proteins, the higher temperatures necessary to make whey cheeses allow that process to take place. When making cheese with whey, you must act fast: the whey must not be more than three hours old. As with regular cheese making, the freshest whey will produce the sweetest cheese. Also note that whey may be warmed over a direct heat source. The techniques used in this chapter are fairly straightforward; see chapter 3 to review them, as needed.

MYSOST,
page 258

Pure Whey Ricotta

With its sweet aroma and utterly delicious, delicate flavor, this cheese is a heavenly treat. You can make it with as much whey as you have; plan for about ½ pound of ricotta per gallon of whey. It's important to use sweet whey that's left over from making hard cheeses only; acidic whey from cheeses coagulated with vinegar, lemon juice, and citric or tartaric acids will not work.

Fresh whey (no more than 3 hours old)

1. Heat the whey in a pot until you see curds precipitating on the top . This usually happens between 190 and 200°F (88–94°C); do not let it boil or the cheese will taste burnt.

2. Turn off the heat; let the whey set for 5 minutes.

3. Gently skim off the curds and place them in a colander lined with butter muslin or an unlined ricotta mold.

4. Let the curds drain for 15 minutes, then refrigerate in a covered container. This ricotta will keep for up to 1 week in the refrigerator.

RICOTTA WITH
WHEY AND MILK

ZIERGERKÄSE,
page 260

RICOTTA WITH WHEY AND MILK

Ricotta comes from Italy; the name means to cook twice or "recook." It is a delicious, soft, fresh curd that can be eaten on its own or perhaps drizzled with honey for a delectable dessert. Traditional versions are made from ewe's-milk whey, but this recipe uses cow's milk for those of us who don't have a milking sheep in the backyard. It's important to use sweet whey that's left over from making hard cheeses only; acidic whey from cheeses coagulated with vinegar, lemon juice, and citric or tartaric acids will not work.

2 gallons fresh whey (no more than 3 hours old)

1 quart whole milk, for increased yield (optional)

1½ teaspoons citric acid dissolved in ½ cup cool water

1 teaspoon cheese salt (optional)

Yield: 1–2 cups

1. Pour the whey into a large pot and heat to 160°F (71°C). Add the milk and continue heating to 170°F (77°C).

2. Add the citric acid solution and heat to 185–195°F (85–91°C). When you see white particles of albuminous proteins starting to gather on the surface, stir gently for 30 seconds to consolidate, then turn off the stove and let rest for 10–15 minutes.

3. When the curds have stopped rising to the top, gently ladle them into a ricotta mold and drain. For a fresh ricotta, drain 15–20 minutes; for a richer, more buttery ricotta, drain several hours and chill overnight.

4. Store in a covered container for up to 1 week.

Mysost

This is our variation on the Scandinavian sweet cheese that is often served on toasted bread for breakfast. The color ranges from light to dark brown, depending on the degree of sugar caramelization and whether cream has been added. For a more spreadable consistency, shorten the cooking time. If you would like a sliceable cheese, heat the mixture to a thicker consistency before pouring it into a mold. As a variation, add crushed walnuts to the thickened whey just before it is cooled.

NOTE: Using a wood cookstove is an economical way to make this cheese, because it involves many hours of boiling.

Fresh whey (no more than 3 hours old), from making cheese from 2 gallons of cow's milk

1–2 cups heavy cream (optional)

Yield: 1½ pounds

1. Pour the whey into the pot and heat to 200°F (93°C). Watch the pot carefully for a soft curd layer to appear on the surface. Skim these curds off with a slotted spoon and place them in a bowl. Reserve in the refrigerator. If you don't remove this layer, the whey will boil over.

2. Let the whey boil slowly, uncovered, over low heat. When it has cooked down to 75 percent of its original volume (this can take 6–12 hours), start stirring often to prevent sticking. Add back the reserved curds. At this point, the mixture will begin to thicken.

3. Add the cream, if desired, and stir well. The amount you add will determine the final texture of the cheese. The more cream, the smoother the texture.

4. Pour the mixture into a blender and blend on medium speed for 1 minute or until smooth and creamy. (Hold the top of the blender with a pot holder to prevent the top popping off due to the steam.)

5. Put the mixture back into the pot and continue to cook over low heat, stirring constantly with a rubber spatula. The mixture will continue to thicken.

6. When the mixture approaches a fudgelike consistency, place the pot in a sink full of cold water and continue stirring until it is cool enough to be poured into the mold (traditionally rectangular). If the whey is not stirred while it cools, the cheese may become grainy.

7. Once cool, remove the cheese from the mold, cover, and store in the refrigerator. This cheese will keep for up to 4 weeks in the refrigerator and may be frozen for up to 6 months.

VARIATION: GJETOST

This is the same cheese as Mysost but made with the whey and cream from goat's milk. It tends to be tangier and a little saltier.

ZIERGERKÄSE

You can enjoy this popular German cheese fresh or after aging it for several weeks — it's equally delicious either way. It's important to use sweet whey that's left over from making hard cheeses only; acidic whey from cheeses coagulated with vinegar, lemon juice, and citric or tartaric acids will not work.

2 gallons fresh whey (no more than 3 hours old)

1 quart whole milk, for increased yield (optional)

¼ cup vinegar (I like apple cider vinegar)

1 quart water

1 quart dry red wine

¼ cup cheese salt

Herbs (optional)

Yield: About 1 pound

1. Pour the whey into a large pot. Add the milk, if desired. Heat the mixture to 200°F (93°C).

2. Slowly add the vinegar. Turn off the heat.

3. Allow the mixture to set for 10 minutes. You will see white flakes of protein floating in it.

4. Gently ladle the mixture into a colander lined with butter muslin. When the muslin is cool enough to handle, tie the corners into a knot and hang the bag over the sink to drain for several hours, or until the curds stop dripping whey. Allow the curds to cool.

5. Line a 1-pound cheese mold with butter muslin. Add the curds and press at 20 pounds of pressure for 24 hours.

6. Remove the cheese from the press and gently peel away the cheesecloth.

7. Combine the water, wine, and salt in a large bowl; add the herbs, if desired. Add the cheese and cover. Place the bowl in the refrigerator for 4 days, turning the cheese twice a day.

8. Remove the cheese from the soaking liquid. Drain on paper towels and cover with cheese wrap. The cheese may be eaten fresh or aged for several weeks in the refrigerator before serving. It will keep up to 4 weeks.

10

CULTURED DAIRY PRODUCTS

Many a home cheese maker started off with cultured dairy products such as these. The methods used to produce them are simple and straightforward, and you can use almost any type of milk, including dried milk powder, low-fat milk, whole milk, light cream (25%), and half-and-half, as well as goat's milk or sheep's milk. Just keep in mind that the higher the butterfat content, the higher the yield and the creamier the finished product. What's more, the fresher the milk, the higher the quality of your end result.

Feel free to experiment by adding fresh or dried herbs, a variety of spices, your favorite sweeteners, and other flavor enhancers to any of these dairy products once they are ready.

See chapter 3 to review any of the techniques, as needed.

BUTTER,
page 262

261

BUTTER

Like so many delicious foods, butter was invented out of necessity, specifically as a way of preserving the fat in milk to use during colder seasons. On many family farms, butter is still produced using the time-tested method of agitating cream in a butter churn until the butterfat is released, washed, salted, and eaten. Here is a simple way for you to enjoy making your own butter at home. If you must use ultrapasteurized cream, add one-half packet of buttermilk starter and let the cream set overnight to ripen, then follow the directions below, starting with step 2.

1 pint heavy cream or
whipping cream

Cheese salt (optional)

Yield: About 8 ounces

1. Let the cream set at room temperature (72°F/22°C) for several hours to ripen slightly. Then chill to below 58°F (14°C) before starting.

2. Pour the cream into a 1-quart canning jar with a tight-fitting lid and shake vigorously.

3. After 5–10 minutes, when the butter has formed, pour off the liquid buttermilk and spoon the solids into a bowl.

4. Add ½ cup cold water and press with the back of a spoon to help expel more buttermilk. Pour off the excess liquid and continue adding ½ cup cold water and expelling buttermilk until the liquid runs clear.

5. Add salt to taste, if desired, which will help preserve your butter. Refrigerate for 8–12 hours and enjoy! It will keep in the refrigerator for up to 1 week.

TROUBLESHOOTING

If your butter is not forming, warm up the cream a bit to raise the acidity level. If the butter has a cheesy flavor, it is overly acidic. Sterilize all equipment and next time do not ripen the cream for as long. If your butter is rancid, you used dirty equipment or you didn't wash out all the buttermilk with water. Do not eat butter if it is rancid.

Buttermilk

Unlike traditional or "real" buttermilk, this recipe produces a type of buttermilk that you can use as a starter culture in recipes for soft cheese. The length of setting time alters the viscosity and amount of acidity in the final product; experiment with this for variations in thickness and flavor.

1 quart pasteurized whole or skim milk

1 packet C21 buttermilk starter culture

Yield: 1 quart

1. Heat the milk to 86°F (30°C).

2. Add the starter; cover and let the milk set undisturbed at room temperature (72°F/22°C) for 12 hours, or until coagulated.

3. It is now ready to use, and will keep for up to 1 week stored in the refrigerator.

Recipe from a Home Cheese Maker

Sweet Butter

This rich, sweet butter is fabulous, as is the buttermilk produced in the making. Warm 1 quart of cream to between 90 and 100°F (32–38°C) in a glass canning jar. Add a scant ¼ teaspoon (or half of a small packet) of direct-set mesophilic starter culture, stir, and let set for 8–12 hours. In the morning, chill in the refrigerator until it is 40°F (4°C). Scoop it out and beat in a stand mixer with the paddle attachment at a fairly high speed. (Don't whisk or it will thicken too much.)

As the butter starts to come, reduce the speed of the mixer. Once the butter has come, drain the buttermilk into a measuring cup. This makes between 1 and 1½ cups of buttermilk. Save it to use in baked goods, or just drink it right up — it is so good. (Just don't use it to make cheese.)

Return the same amount of cold water to the butter and beat it again. Drain and repeat several times until it runs clear. At least for the last wash, use water that is not so cold that it makes the butter hard to work, because you now want to press the butter against the sides of the bowl with a wooden spoon to get out the water. Add salt, if desired.

Press the butter into small custard cups, put the cups into small resealable bags, and keep in the refrigerator.

MAKING GHEE (CLARIFIED BUTTER)

Ghee, a highly clarified butter with no milk solids, originated in India. With its high smoking point, ghee is better suited to frying and other high-temperature cooking because it doesn't splatter or burn as easily as butter, and it imparts a depth of flavor. In Europe, during the Middle Ages, butter was always clarified for cooking.

Ghee is incredibly easy to make: Gently melt a quantity of butter in a heavy saucepan over low heat. Allow it to bubble for several minutes, then remove from the heat.

Let the butter set for several minutes to allow the milk solids to settle to the bottom of the pot. Skim off the butterfat from the surface, then pour the clear yellow liquid through two layers of butter muslin.

It will solidify when chilled. Store in the refrigerator for up to a year.

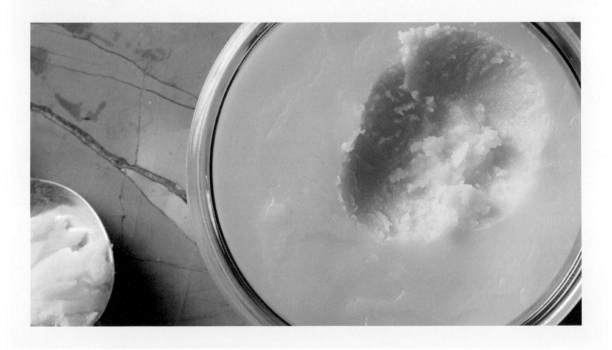

SOUR CREAM

In the old days, all you had to do to create sour cream was leave milk on the back of the woodstove to sour. Today, because of the high-heat treatments that store-bought milk goes through, all you would get is spoiled milk, which is where this recipe comes in. Sour cream's tangy taste may be made more or less acidic with the length of setting time. Longer setting produces a tangier product. For a low-calorie sour cream, substitute low-fat milk for the light cream.

1 quart pasteurized light cream or half-and-half (25–40% butterfat)

1 packet direct-set sour cream starter culture

Yield: 1 quart

1. Heat the cream to 86°F (30°C).

2. Add the starter; cover and let the cream set undisturbed at room temperature (72°F/22°C) for 12 hours, or until coagulated.

3. The sour cream is now ready to use, and will keep for up to 1 week stored in the refrigerator.

CUP CHEESE

Cup cheese, or soda cheese, is an old Amish and Mennonite recipe still made in Pennsylvania. The curdled milk must be made fresh from the animal; store-bought sour milk will not work.

1 gallon sour milk (let stand at room temperature until thickened)

3 tablespoons unsalted butter

½ teaspoon baking soda

1 cup cream

1 egg, beaten

1 teaspoon cheese salt

YIELD: 1 POUND

1. Heat the milk to 115°F (46°C). With a knife, cut both ways through the curdling milk.

2. Line a colander with butter muslin. Pour in the milk. Tie the corners of the muslin into a knot, hang up the bag, and let drain for 8–12 hours.

3. Crumble the curds and stir in the butter and baking soda. Let set for 5 hours.

4. Place the mixture in a double boiler and heat until it melts. Add the cream and stir until smooth.

5. Add the egg and salt, and bring the mixture to a boil. Pour into dishes. Serve with homemade bread and butter.

PASHKA

This is one version of this cheese, which is also known as Russian Easter cheese. Pashka is super rich and served in small portions as a holiday dessert. Traditional recipes call for making pashka in a 4- to 5-inch-diameter clay flowerpot (clean and sterilized), which has a convenient hole for draining, but you can make your own mold by using a large yogurt container with a hole in the bottom or just use a fine-mesh sieve.

5½ cups cream cheese or cottage cheese

2 cups plain yogurt or sour cream

½–1 cup (1–2 sticks) unsalted butter, softened (1 stick is enough if using homemade cheese)

5 eggs, beaten

2 cups sugar (less 1–2 tablespoons if using homemade cheese)

1–2 tablespoons finely chopped, blanched, and skinned almonds

1 tablespoon currants

1 teaspoon grated lemon peel

Vanilla extract

Yield: Ten 4-ounce servings

1. Push the cream cheese through a sieve into a bowl. Add the yogurt, butter, and eggs, and stir with a fork. Pour into a saucepan.

2. Heat the mixture to 90°F (32°C), stirring constantly. Hold at this temperature until the mixture starts steaming. Do not allow it to boil.

3. Remove from the pan, place in a bowl set in a larger bowl of ice, and chill the mixture. Add the sugar, almonds, currants, lemon peel, and vanilla to taste. The mixture will be very sweet at this point, but a good deal of sugar will drain out.

4. Line a pashka mold or small flower pot with a large piece of butter muslin. Pack in the mixture and neatly fold the muslin over it. Place a 3–5 pound weight on top.

5. Stand the pot in a dish to catch the whey. Set in a cool place for 12–14 hours.

6. Invert onto a serving dish and remove the cloth. Spoon into individual dessert dishes. If not serving immediately, place the dessert dishes on a tray, cover the tray with dampened cheesecloth, and refrigerate for up to 5 days.

CRÈME FRAÎCHE

If you've always loved the rich, tangy taste of crème fraîche but not the usual high cost, here's your answer: Make it yourself, and in very short order. It's wonderful as is, dolloped over desserts or fruit (it has a particular affinity for grilled peaches and is even creamier when drained and used as a base for dips and spreads or stuffed into fresh figs or dried dates).

1 quart pasteurized light cream or half-and-half (25–40% butterfat)

1 packet crème fraîche starter culture

Yield: About 1 pound

1. Heat the cream to 86°F (30°C).

2. Add the starter; cover and let the cream set undisturbed at room temperature (72°F/22°C) for 12 hours, or until coagulated.

3. It is now ready to use, and will keep for up to 1 week stored in the refrigerator.

CRÈME FRAÎCHE

BUTTERMILK,
page 263

BUTTER,
page 262

YOGURT,
page 272

269

A CHEESE MAKER'S STORY

Topher Sabot
CRICKET CREEK FARM
Williamstown, Massachusetts

WHEN DICK AND JUDE SABOT heard that Cricket Creek, which abutted their own property and was one of the oldest dairy farms in Massachusetts, had shut down, they decided to buy it themselves, mainly to protect the land. They then came up with the idea of running a small-scale dairy farm, adding value to the milk through cheese. This was back in 2001, when Topher had graduated from Williams College (where his father was a professor) and moved out west. "The plan was never for them to do it all themselves, but things took a very different path when my father died suddenly in 2005."

Thanks to a tremendous outpouring of support from the local and farming communities, Topher and Jude were able to realize Dick's original vision of a farmstead creamery within a couple years. By 2009, it was clear they had to handle the day-to-day operations, with Topher managing the herd and farm and Jude overseeing the cheese making — and both learning the skills on the job. "That significant step precipitated the next phase of the farm that led us to where it is today."

Today, Cricket Creek produces about 30,000 pounds of cheese a year from milk produced by 30 to 40 cows, who are walked through their pastures twice a day, every day (weather and season permitting). "We make cheese five days a week at the high point and will never drop down below three days a week. There's no such thing as an off-season."

Topher says they are strategic in choosing the types of cheese they make, including Maggie's Round, an ever-popular Italian-style Tomme with a long shelf life. "Maggie's Round is ready anywhere from two to eight months of aging, so we always have cheese to sell. When it's aged out over a year, we sell it as Maggie's Reserve. Having those options is really important." Making fresh cheeses, which yield more cheese per pound of milk than aged ones, also helps buttress their bottom line.

Topher feels a big responsibility as a steward of this historical spot. "It's an amazing feeling because it's operating on a scale that goes beyond the human lifespan. To think about all the people who cared for it in the past and then those who will tend to it in the future. Hopefully there will be a working dairy farm here for generations to come."

DEVONSHIRE CLOTTED CREAM

Clotted cream was developed in southern England as a way to make delicious use of an abundance of rich, creamy milk during the summer. Then, a pot of milk would be left on a hot stove overnight to allow the cream to rise to the surface, leaving a layer of thick cream covered by a wrinkled skin. This updated method uses the same idea of slow, constant heat to thicken the cream and encourage slight ripening by bacteria. This clotted cream also has a distinctly golden brown color and nutty flavor, thanks to caramelized milk sugar. In other words, it's decidedly delectable.

1 quart heavy cream
(48–50% butterfat)

Yield: About 1 cup

1. Slowly heat the cream to 175°F (79°C) in a double boiler over medium heat (it should take about 30 minutes). Stir occasionally to ensure even heating.

2. Without further stirring, raise the temperature to 180–190°F (82–88°C). Maintain that temperature for 45 minutes. The surface will develop a thick, wrinkled appearance.

3. Remove from the heat and cool rapidly in a pan of ice water. Cover and place in the refrigerator for 12–24 hours. Skim the clotted cream off the top with a ladle; discard the remaining liquid. Store in a covered container in the refrigerator for 1 week.

Yogurt

Ever-popular yogurt has been eaten since ancient times, when it spread from the Arab nations and throughout the Middle East to central Asia and southern Europe. It's one of the easiest — and most versatile — dairy products to make and to have on hand.

2 ice cubes

1 quart milk (any type)

1 packet yogurt starter culture or 2 tablespoons yogurt with live cultures

Yield: 1 quart

1. Roll the ice cubes around the pot until it feels cold to the touch from the outside (see page 70). Add the milk.

2. Heat the milk to 185°F (85°C).

3. Let the milk cool to 116°F (47°C). Add the starter and mix well.

4. Cover and let set at 110–116°F (43–47°C) for at least 6 hours, or until set to the desired consistency.

5. Store in a covered container in the refrigerator for up to 2 weeks.

NOTE: You can use store-bought yogurt with live cultures in place of powdered yogurt culture, but be aware that it may contain additives and the culture may not be as strong as that made from the powdered culture.

Homemade yogurt can be used to make a new batch 8–10 times. When it stops working, start a new batch with the culture powder.

FOR THICKER YOGURT

Yogurt will thicken as it cools, but if you want an even thicker consistency, you can add one of the following to each quart of milk:

¼ cup dry milk powder

1 tablespoon of a thickener such as carrageenan, pectin, or gelatin

Goat's milk produces a thinner yogurt than cow's milk. To thicken it, mix 1 drop of rennet in 4 tablespoons of cool, nonchlorinated water, then add 1 tablespoon of the diluted rennet to the milk.

MAKING LABNEH

Labneh (also laban, labne, lebnah, or lebni) is a Middle Eastern specialty made by straining yogurt until extra thick, like a thicker Greek yogurt in some cases or, when strained even further, a soft cream cheese that can be rolled into balls (see below). Labneh is traditionally made with cow's milk and strained through a brown paper bag, but you can also use goat's or sheep's milk, and muslin or cheesecloth for draining. For a savory breakfast or snack, drizzle labneh with olive oil, sprinkle with chopped pitted olives, dust with za'atar, and serve with toasted pita triangles.

For a tasty appetizer, try this Syrian dish known as *laban dahareej*. Roll the cheese into balls, make a depression with the back of a knife, put in a drop of olive oil, and sprinkle with chopped mint; serve with toothpicks. For a different take, first place the cheese balls on a tray or serving dish and allow to set for 8–12 hours in the refrigerator until firm. Then place in a glass jar, cover with olive oil, secure with a tight-fitting lid, and refrigerate. It will keep for several weeks.

To serve, arrange cheese balls in a small dish, spoon a small amount of oil over the balls and sprinkle with chopped fresh mint and freshly ground black pepper. The cheese will be soft enough to spread over slices of Syrian bread or sesame flatbread.

KEFIR FROM LIVE GRAINS

With its effervescence (and trace amount of alcohol), kefir is often referred to as the "champagne of milk." It has been enjoyed in Russia for thousands of years, where the name means "good feeling" or "pleasure," and the mountain people believed it had healing properties. This recipe follows tradition and uses live grains. With proper care, you will be able to use the grains to make kefir for many, many years. Please get your grains from a reliable source.

1 container (about 3 tablespoons) kefir grains

3 cups fresh pasteurized milk (whole or low-fat)

Yield: 3–4 cups

1. Place the grains in a 1-quart glass jar and pour the milk over them.

2. Cover the jar loosely and let set at room temperature for 24 hours, agitating the jar from time to time.

3. After 24 hours, gently stir the mixture with a spoon.

4. Place a nonmetal strainer over a bowl and pour the kefir through it. Gently stir until most of the kefir falls through into the bowl. Refrigerate and enjoy!

5. To save the grains for next time, do not rinse them. Leave some curds clinging to them, and do not try to remove every last drop of kefir. Gently scoop the grains with the clinging curds into a clean glass jar. In a week or two, you will notice that they have expanded.

6. For your next batch of kefir, pour 3–4 cups of fresh milk over 3 tablespoons of grains and repeat steps 1 through 5.

TROUBLESHOOTING

Depending on how mild or strong you like your kefir, pour more or less milk over the grains. Experiment by using anywhere between a 5:1 ratio of milk to kefir grains (2 cups of milk to 3 tablespoons of grains) and a 25:1 ratio (10 cups of milk to 3 tablespoons of grains). When you have too many grains to use up, slow kefir production by culturing it in the refrigerator, straining the grains once a week.

KEFIR FROM STARTER CULTURE

This recipe uses a powdered form of kefir starter and is slightly less fermented than the kefir made from live grains.

1 quart pasteurized whole milk

1 packet kefir starter culture

Yield: 1 quart

1. Heat the milk to 86°F (30°C).

2. Add the starter; cover and let the milk set undisturbed at room temperature (72°F/22°C) for 12–15 hours, or until coagulated.

3. It is now ready to use, and will keep for up to 1 week stored in the refrigerator.

KEFIR CHEESE

Like yogurt, kefir can be drained to make a thick, spreadable cheese. Using homemade kefir to make this cheese is preferable, as it produces a much thicker curd than does commercial kefir.

1 quart fresh kefir

Cheese salt (optional)

YIELD: ABOUT 8 OUNCES

1. Let the kefir come to room temperature (72°F/22°C).

2. Pour the kefir into a colander lined with butter muslin. Tie the corners of the muslin into a knot and hang the bag to drain for 12–24 hours, or until the cheese has stopped draining and has reached the desired consistency.

3. Remove the cheese from the bag and place in a bowl. Add salt to taste, if desired. Cover the bowl and store in the refrigerator for 1–2 weeks.

Jeremy Logo
THE PANTRY FINE CHEESE
Ontario, Canada

A FORMALLY TRAINED SOMMELIER AND FROMAGER (and home cheese-making hobbyist), Jeremy Logo realized a lifelong dream when he opened The Pantry Fine Cheese in Toronto with his wife, Kathryn, in 2008. His sole mission was to showcase only Canadian cheese and only *artisanal* cheese at that, stocking about 75 cheeses from some 30 different producers. (The only non-Canadian cheeses he sells are Parmigiano-Reggiano and Pecorino from Italy, which he says simply can't be replicated properly.)

When deciding whether to offer a new cheese in his shop, he considers the quality of the product, but also the dedication of the cheese maker (and the farmer, when they're not one and the same). "I only work with cheese makers who know where their milk comes from, so I know I'm getting good-quality products and that the animals have a good quality of life." The process is also key, which is why he sticks with small-scale operations and looks for innovative methods, such as aging cheeses with cedar to create a singular flavor.

"Consistency is another huge factor for me, and that's easier for new cheese makers to control when they are focusing on just a couple of cheeses." He said he had to stop selling one cheese after it kept changing with each batch — customers were just too confused.

Then there's the ever-important matter of pricing. Since opening his shop, he has definitely noticed that people want to shop locally and are willing to pay more for known products, but he still has to remain competitive. That has meant working with some cheese producers to reach a mutually agreeable price point.

Because there aren't many others who concentrate on Canadian products, Jeremy would like to open a few more locations and continue to spread the word about the quality of Canadian cheese. "I live in a neighborhood where there are lots of young families and I love it when kids come in and say 'I want Five Brothers cheese!' or 'I love Grey Owl!' It's great to see even the youngest folks get interested in it, too." Because for Jeremy, cheese is more than just something delicious, it's a lifestyle. "Cheese making is not profitable. It's a very tough business. A lot of time goes into just one wheel of cheese."

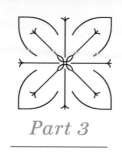

Part 3

FOR THE LOVE OF CHEESE

BUILDING A BEAUTIFUL CHEESE PLATE

hen it comes to sharing your handiwork with friends and family, do not underestimate the value of proper cutting techniques and serving temperature so as to realize the full potential of your cheese efforts. There's nothing wrong with putting out simple slices or cubes for a casual gathering, but done correctly, portioned cheese will not only look its best, it will also taste its best — and stay that way when stored.

Most experts agree that an outstanding cheese will speak for itself, but you can help showcase your artistry by following some general guidelines for pairing cheeses with complementary and contrasting tastes and textures, as well as selecting the right accompaniments and beverages.

Cutting Tips and Techniques

A cheese sample's flavor and texture will vary depending on what part of the cheese it was taken from — for example, its proximity to the rind (if any) — or how much veining is present (as in blue cheese). Correctly cutting a whole cheese aims to provide a uniform cross-section for the most uniform enjoyment. Here are some guidelines for cutting each type of cheese so it presents its full flavor potential.

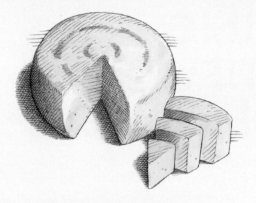

Large, wheel-shaped hard cheeses
Cut into triangular wedges from the center out, then cut the larger pieces again crosswise if desired.

Cylinder-shaped cheeses
Cut large cylinders crosswise into large rounds, then cut the rounds into triangular wedges. Smaller cheeses can be cut into rounds; very small cheeses can be left whole or cut in half.

Pyramid-shaped cheeses
Cut into quarters or smaller wedges.

Large, wheel-shaped soft cheeses
Cut the wheels into triangular wedges from the center out, then cut the wedges lengthwise into narrower wedges.

Square or rectangular soft or buttery cheeses

Cut rectangular pieces from the center of the cheese so that after serving, you can push the two cut faces back together to contain the runny center and prevent it from oozing out of the crust. If the entire cheese will be eaten at once, or if the center is not runny, cut the cheese into triangular wedges.

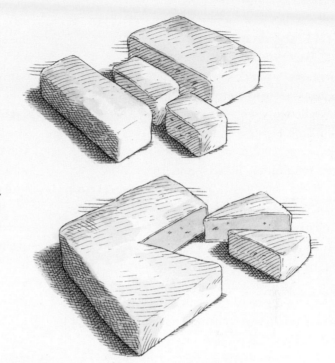

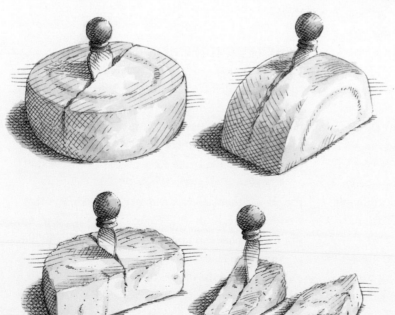

A whole Parmesan wheel

This is an art all to itself: Use a special almond-shaped knife to preserve the cheese's characteristic granular quality. First split the cheese in half by making ½- to ¾-inch-deep cuts across the diameter and down the sides. Then insert two knives into the cuts and pull them in opposite directions; the knives will act as levers to split apart the cheese. Repeat this procedure to divide the cheese into smaller portions, always trying to keep a consistent ratio of cheese to rind.

Putting Together a Cheese Board

There's no "wrong" way to assemble a cheese platter — all who sample it are guaranteed to be enthralled by your accomplishments. That said, there's most assuredly a "right" way (or two or three, depending on who you ask). Here are some tips from John Antonelli (see profile, page 287).

DON'T BE FUSSY ABOUT THE PRESENTATION. If you're having a small gathering of close friends, by all means, pop the cheeses on a board in their natural glory and let everyone admire before helping themselves. Or, for neater nibbling, cut the cheeses into large chunks. For larger parties, it's easier for guests if you cut hard and semi-soft cheeses into proper portions, but you can still opt to leave soft cheeses whole.

DON'T OVERWHELM WITH TOO MANY CHEESES. Include no more than three to five types, varying in texture, color, style, age, and milk type (think cow, goat, and sheep). Having a sensible selection (see below) will also ensure there's a cheese to please each person. Plan on 1 to 1½ ounces of each type of cheese per person if the cheese board precedes a meal; otherwise double the amount.

CONSIDER THE CONDIMENTS. These can include fresh or dried fruit; chutneys, preserves, or honey; pickles and cured meats. Strive for a balance of sweet, salty, and savory to bring out the different flavors of the cheeses. Bread and crackers are, of course, essential.

DON'T OVERTHINK THE DRINK PAIRINGS. When in doubt, go with bubbles. Sparkling wine, beer, cider, or even a good mineral water— their effervescence cuts through the richness of the cheese, cleansing the palate and priming it for the next scrumptious bite. (For more detailed advice, see What to Pair with Cheese: The Age-Old Conundrum, page 294.)

CHEESE SELECTION

The first step is to determine your group size and also the style of the gathering (casual get-together with close friends versus more formal holiday party with a mix of guests).

When serving a cheese board to close friends and acquaintances, it's easy to tailor your selection of cheeses to match your guests' preferences — and they're probably more willing to give something new (and stinky!) a try if you're the one who actually made it.

When you're not so familiar with the folks on your guest list, or if it's a more formal party, you may prefer to stick with safer, universally adored cheeses.

GENERAL GUIDELINES

TEMPERATURE. Let the cheese sit at room temperature for a while before serving, leaving it in its wrapping so it doesn't dry out. Too-cold cheese won't have the depth of flavor; too-warm cheese loses some (or all) of its integrity. Plan on about 90 minutes per pound to reach 72 degrees (the sweet spot).

SERVING VESSELS. If guests will be cutting the cheese themselves, arrange cheeses on a sturdy wooden, slate, or marble board; otherwise, you can put precut cheese on pretty much anything you like.

KNIVES. Have one knife for each cheese you are serving to keep the flavors distinct. Any kind of knife will do, but special cheese knives are worth collecting. Hard cheeses require sharp knives or cheese planes, while soft cheeses call for gentler spreaders.

FOR SMALL GROUPS (2–10 GUESTS)

3 or 4 cheeses

Leave most cheeses in chunks for self-portioning.
Always cut one piece to set the example, unless
it's too runny (leave it whole) or too hard (cut into
proper portions as a courtesy to your guests).

FOR LARGE PARTIES (10+ GUESTS)

5 or 6 cheeses

To ensure that everyone gets to try every cheese, cut them into proper portions, leaving soft cheeses whole so guests can dive right in. Try to keep the integrity of the cheese intact by mimicking the original shape (i.e., no cubes with toothpicks).

FOR A FORMAL DINNER PARTY (ANY NUMBER OF GUESTS)

3 cheeses

Preportion the cheese and place one of each type on an individual plate. This can be served as an appetizer or dessert.

ALTHOUGH ANTONELLI'S CHEESE HOUSE only opened their doors in 2010, the business has become an essential part of the highly regarded Austin food scene as well as the American cheese-making community. So much so that John Antonelli — who founded the shop with his wife, Kendall, practically on their honeymoon — was appointed president of the board of directors of the American Cheese Society (ACS) in 2017. "I'm eager to volunteer even more time to the ACS, which is near and dear to my heart," says John.

After speaking with John about the past, present, and future of the American artisanal cheese industry, it's easy to see why he was chosen to lead during such a critical time of growth and challenge (brought on by new food-safety requirements and increased competition, among other pressing matters).

"People want to know where their food is coming from, and that's where we have an advantage. Our cheese makers are making cheese that has a very limited number of ingredients and the quality is high, so I think that consumers are just going to come around to it. Artisanal cheese won't be just for holidays and special occasions but for brunch or a salad. We'll start to see that more and more. We've had a 10 percent growth this year at ACS and I don't see that slowing down. We're a young industry that has a long way to go to hit the pinnacle.

"American cheese makers are incredibly innovative and coming up with so many unique ways to tweak the basic techniques — of which there are so few — to create entirely new flavor profiles. We've only seen the beginning. It has and will continue to be a veritable explosion within the industry."

John Antonelli
ANTONELLI'S
CHEESE HOUSE
Austin, Texas

INDISPENSABLE ACCOUTREMENTS

Now that you've made your own cheeses, preparing the sidekicks will be a cinch. (And if you buy some or all of the cheese — no judgment! — you can at least prepare these homemade accompaniments.) Mix and match any of the following to provide contrasting tastes and textures to your cheese board.

1. TARRAGON MUSTARD
2. MEMBRILLO
3. PORTER MUSTARD
4. HOT PEPPER JELLY
5. LAVASH FLATBREAD
6. HONEY MUSTARD
7. OATCAKES
8. TOASTED PITA CHIPS

Green Tomato Jam

1½ pounds green tomatoes, cored and chopped (about 3½ cups)

2½ cups sugar

Zest and juice of 1 lemon (about 3 tablespoons juice)

1 cinnamon stick

1 teaspoon ground cardamom

1 slice (¼ inch thick) peeled fresh ginger

1. Mix the tomatoes and sugar in a medium bowl, stirring well to distribute the sugar. Let stand, covered, for at least 4 hours or up to overnight.

2. Combine the tomato mixture, lemon zest and juice, cinnamon, cardamom, and ginger in a heavy-bottomed pot. Bring to a simmer and cook, stirring occasionally, until tomatoes are translucent and the juice has started to gel and thicken, 15–20 minutes.

3. Remove the cinnamon stick and ginger. Transfer the jam to clean jars and let cool. Screw on lids and store in the refrigerator for up to 1 month, or freeze for up to 3 months.

NOTE: The level of sugar in this recipe makes it unsuitable for boiling-water canning.

Fresh Fig Chutney

MAKES ABOUT 2 CUPS

2 tablespoons extra-virgin olive oil

3 shallots, finely diced

½–¾ cup packed dark brown sugar

½ cup apple cider vinegar

Zest and juice of 1 lemon (about 3 tablespoons juice)

¾ cup golden raisins

½ cup dried figs, stemmed and chopped

1 (½-inch) piece fresh ginger, peeled and minced

1 teaspoon mustard seeds

1 cinnamon stick

½ teaspoon whole allspice

½ teaspoon kosher salt

1 pound fresh figs, stemmed and chopped

1. Heat the oil in a medium saucepan over medium-high heat. Add the shallots and cook, stirring occasionally, until translucent and starting to turn golden, about 10 minutes.

2. Add ½ cup brown sugar, the vinegar, lemon zest and juice, raisins, dried figs, ginger, mustard seeds, cinnamon, allspice, and salt to the pan.

3. Bring to a simmer over medium-low heat and cook until raisins and dried figs are tender, 15–20 minutes.

4. Add fresh figs and cook for 20 minutes longer, or until the mixture thickens. Taste and add up to ¼ cup more of brown sugar as desired.

5. Remove the cinnamon stick. Transfer the chutney to clean jars and let cool. Screw on lids and store in the refrigerator for up to 1 month, or freeze for up to 3 months.

NOTE: The level of sugar in this recipe makes it unsuitable for boiling-water canning.

Membrillo
(Quince Paste)

MAKES ABOUT 2 CUPS

2 pounds ripe quince, peeled, cored, and cut into wedges (about 4 cups)

Juice of 1 lemon (about 3 tablespoons)

2–3 cups sugar

⅛ teaspoon kosher salt

1. Combine the quince and lemon juice in a medium stockpot. Cover fruit with water. Bring to a boil and cook until quince is very soft, about 30 minutes.

2. Drain quince in a sieve or colander, and let cool for 10 minutes.

3. Purée the quince until very smooth in a blender or food processor and pour into a measuring cup, noting the volume.

4. Combine the quince, an equal amount of sugar, and the salt in a heavy-bottomed pot. (For example, if there are 2 cups of purée, add 2 cups of sugar.) Place pot over medium-low heat and simmer, stirring occasionally, until a thick paste begins to form, 1½–2 hours.

5. Preheat the oven to 150°F (66°C), or the lowest setting possible. Line a 9-by-13-inch baking dish with parchment.

6. Spread the thick quince paste onto the parchment. Bake for 90 minutes, or until the surface of the membrillo is just barely tacky. Let cool completely before storing, covered, in the refrigerator for up to 2 months.

Hot Pepper Jelly

MAKES ABOUT 2 CUPS

2 pounds tart apples such as Granny Smith, chopped (skins, cores, and all; about 6 cups)

3–6 jalapeños (depending on heat preference), cut into thick rings (seeds removed if less heat is desired)

1 large red bell pepper, ribs and seeds removed, chopped

1½ cups white vinegar

1¾ cups sugar (less or more depending on juice yield)

1. Combine the apple chunks, jalapeños, bell pepper, vinegar, and 1½ cups water in a large saucepan. Bring to a boil, then reduce heat and simmer, stirring occasionally, until apples and pepper are very soft, 15–20 minutes.

2. Line a large sieve with a square of doubled cheesecloth and set it over a medium bowl.

3. Use a potato masher or whisk to break the apples up into a pulp. Pour the contents of the pot into the prepared sieve. Let the pulp drain for at least 4 hours or up to overnight. If juices do not drain from the pulp, return pulp to the pot, add ½ cup more water, and cook for 5 minutes longer. Pour into the sieve and continue, pressing on solids to extract as much liquid as possible.

4. Measure the juice extracted from the pulp: There should be about 2 cups. Pour the juice into a 10-inch skillet and add the sugar.

5. Over medium-high heat, bring the juice and sugar to a boil, stirring to dissolve sugar. As the juice boils, skim off any scum that forms on the surface. The jelly will gel at about 220°F (104°C), or when the jelly drips in a sheet off a metal spoon.

6. Pour the jelly into clean jars and let cool. Screw on lids and store in the refrigerator for up to 4 months.

NOTE: The level of sugar in this recipe makes it unsuitable for boiling-water canning.

QUICK-PICKLED CARROTS

MAKES 1 PINT

1 cup apple cider vinegar

1 cup water

1½ teaspoons kosher salt

8 ounces carrots, cut into 1½-inch–long sticks

1–2 dried hot chiles

½ teaspoon celery seed

¼ teaspoon whole black peppercorns

1. Bring the vinegar, water, and salt to a boil in a small saucepan. Pack the carrots, chiles, celery seed, and peppercorns into a pint jar. Pour the vinegar mixture over the carrots and screw on the lid.

2. Let the carrots pickle in the jar at room temperature for a day or longer before eating. Store for up to 1 month in the refrigerator.

INFUSED HONEYS

Honey and cheese are a natural pairing, and infused honeys can add even more intrigue to that relationship. These delectable drizzlers are easy to make: Simply drop 1–2 tablespoons of dried herbs or other flavorings into a clean jar and cover with about 1 cup of mild-flavored honey. Let the flavors infuse for 5–7 days. Flip the jar over once or twice during this period to make sure the herbs float throughout the honey. Strain the honey through a sieve to remove the herbs.

Flavors to try include the following.

Oregano and hot dried chiles

Lavender flowers

Cinnamon stick, star anise, and dried ginger

Thyme (especially lemon thyme) and black peppercorns

Sage and rosemary

Dried orange peel, allspice berries, and cardamom pods

NOTE: It's best to use dried ingredients in infusions, as honey provides an ideal anaerobic environment for the growth of botulism toxin.

Flavored Mustards

Homemade mustards are a treat in all respects: easy to make, great for gift-giving, and fantastic to eat. The basic formula for mustard is simple: twice as much liquid — beer, wine, vinegar, water, or any combination of these — as mustard seeds, plus other flavorings.

Use the ingredient lists below as inspiration for mustard experimentation. Note that mustards lose their heat over time, so it's much better to make smaller batches more frequently than to make one sizable batch.

EACH MAKES ABOUT 1½ CUPS

Tarragon Mustard

½ cup yellow mustard seeds

½ cup white wine vinegar

½ cup white wine

1 tablespoon fresh tarragon leaves

½ teaspoon kosher salt

½ teaspoon sugar

Porter Mustard

½ cup brown mustard seeds

1 cup porter or stout beer

1 tablespoon molasses

½ teaspoon kosher salt

Honey Mustard

¼ cup brown mustard seeds

¼ cup yellow mustard seeds

1 cup apple cider vinegar

2–3 tablespoons honey, to taste

½ teaspoon kosher salt

1. Soak mustard seeds in specified liquid overnight.

2. Purée seed mixture in a blender or food processor with remaining ingredients until desired consistency.

Oatcakes

MAKES 18

1½ cups thick-cut rolled oats

1½ cups all-purpose flour, sifted, plus more for dusting

⅓ cup sugar

½ teaspoon baking soda

¼ teaspoon kosher salt

8 tablespoons (1 stick) cold unsalted butter, cut into ½-inch cubes

¼ cup buttermilk

1. Preheat the oven to 350°F (180°C). Line two baking sheets with parchment paper.

2. Combine the oats, 1 cup of the flour, the sugar, baking soda, and salt in a large bowl. Scatter butter over the mixture. Using a pastry blender or two dinner knives, cut in the butter until it resembles coarse crumbs. Stir in the buttermilk to make a moist dough.

3. Transfer the dough to a lightly floured work surface. Using the remaining ½ cup flour, roll the dough out to ⅜ inch thick. Use a 2½-inch cutter to cut out rounds. Gently transfer each round to a baking sheet. Gather the scraps, reroll, and cut out more rounds.

4. Bake for 10–15 minutes, or until oatcakes are golden-edged and crisp. Place the baking sheets on wire racks and let oatcakes cool completely. Store for 2–3 days in an airtight container at room temperature.

LAVASH FLATBREAD

MAKES 4

6 cups all-purpose flour, plus more for dusting

2 teaspoons kosher salt

1¾ cups water

¼ cup plus 1 tablespoon extra-virgin olive oil

1. Preheat the oven to 400°F (200°C). Combine 5½ cups of the flour and 1½ teaspoons of the salt in a large bowl. Stir in the water and ¼ cup of the oil to make a stiff dough.

2. Transfer the dough to a lightly floured work surface. Using some of the remaining ½ cup flour, knead until the dough just comes together. Cover the dough with the overturned mixing bowl and let it rest for 30 minutes.

3. Divide the dough into four balls. Using the rest of the flour, roll each dough ball into a round, making it as thin and crackerlike as possible. Transfer each disk to a baking sheet as it is rolled out. Drizzle with some of the remaining 1 tablespoon oil and the remaining ½ teaspoon salt.

4. Bake (in batches, if needed) for 15 minutes, or until the lavash is crisp and browned. Let cool completely on a wire rack. Store for up to 3 days in an airtight container at room temperature.

TOASTED PITA CHIPS

MAKES 32

4 pita breads (6-inch size)

3 tablespoons extra-virgin olive oil

½ teaspoon kosher salt

½ teaspoon freshly ground black pepper

½ teaspoon ground sumac (optional)

½ teaspoon dried oregano (optional)

1. Preheat the oven to 400°F (200°C). Cut each pita bread into eight wedges. Gently pry each wedge apart to make two triangles. Put the pita pieces in a large bowl.

2. Drizzle the oil over the pieces and toss gently to coat. Sprinkle with the salt and pepper, then with sumac and oregano, if desired. Toss again to distribute the seasonings.

3. Spread the pita pieces evenly over two baking sheets. Bake for 7–10 minutes, or until pita chips are crisp and brown. Rotate the sheets halfway through the baking time if the chips are browning unevenly. Let cool on the baking sheets. Store for 2–3 days in an airtight container at room temperature.

What to Pair with Cheese: The Age-Old Conundrum

Think fast: What's the first spirit that springs to mind when you think of serving cheese — especially of your own making? Odds are it's a lusty red wine, long the chosen companion. Well, think again. According to Steve Jones, owner of Cheese Bar in Portland, Oregon, not to mention myriad other experts, red wine — or any wine — is often *not* the best choice. "Cider is the secret weapon. It has all the magic bullets that work well with cheese: acidity, fruit, and effervescence (sometimes)."

Steve should know. In 2005, he opened a small (350-square-foot) shop in Portland that was completely focused on cheese. Five years later he decided to bring beverages into the equation and opened Cheese Bar, which offers about 250 cheeses, 50 wines, 50 beers, 20 hard ciders, and a menu that's "cheese-focused but not cheese-obsessed." It's also very seasonal. "We do focus on pairings but we really try and have fun with it. Cheese at the end of the day is peasant food and we don't want to turn it into this serious thing."

His main goal is to deal with the smallest producers he can find, what he refers to as "the little bitty guys," whether of cheese, beer, or wine. Not that he's exclusively local. "I can't ignore the rest of the world. Parmigiano-Reggiano is only made in Italy." What makes his spot such a standout are the different cheese boards diners can indulge in while there. "The boards let us keep the inventory moving and allow us to buy fun, funky stuff that people want to take home."

Here are the many things Steve has learned in his many years of selecting, selling, serving, and pairing cheese in all its glory.

General Pairing Tips

When it comes to pairing cheese with spirits, Steve explains that there are three ways to go about it: comparative, contrasting, or regional. "With a good pairing it becomes 'that's a nice addition to this flavor,' but with a great pairing the two flavors come together into an even better third flavor that's all the more elevated." He also says you learn more from mistakes than successes, so keep an open mind and take copious notes. And you can start with either the cheese or the beverage, but cheese is often the easiest way to go.

COMPARATIVE. This involves pairing like flavors. For example, a beer with a little malty sweetness goes nicely with a Gouda that has some of that same sweetness, so you are playing sweet on sweet.

CONTRASTING. Try putting a big, salty Stilton up against a syrupy sweet port. The striking contrast is where the magic happens in this classic meet-up.

REGIONAL. What grows together goes together. Pairing a Norman cider and Camembert can end all arguments right there, they work so well together, as do English cheddar and Somerset cider or a pairing from the same region of Vermont. They're coming from the same soil, so it makes sense.

Pairing with Multiple Varieties of Cheese

If you're serving a bunch of cheeses, then put out a beer, a cider, and a wine and let people play with those. This is the reason to go with a regional cheese board. "That's always my favorite thing to do," says Steve. "It's always such an aha moment for people."

When picking just one wine, go for a rosé, because it splits the difference between red and white. A sparkling rosé is better yet — effervescence cuts through the fats and also the tannins.

For a gradation tasting or "flight," let the cheese lead and start with the simplest and mildest varieties and work up to the most complex. Pair the appropriate beverages with those.

Pairing with Specific Types of Cheese

BLOOMY RIND CHEESES, with their base note of earthy mushroom, want to pair with something funky. Serve them with Norman ciders, which are dirtier than American ciders, and funky beers — a.k.a. farmhouse, wild, saison, or sour. On the other hand, sparkling wine will cut through the cheese's funk.

FRESH GOAT CHEESES pair incredibly well with rosé for wine and Gose, a German-style beer that is a touch tart and has a little more salt than usual. "This is one of my all-time favorite comparative flavorings." Goat cheese also goes really well with Sauvignon Blanc, Sancerre, Grüner Veltliner, and other classic white wines.

SMOKED CHEESE (scamorza) works nicely with light-bodied beers that have a malty sweetness.

FRESH CHEESES go great with red table wine, Chianti, and simpler beers like pilsner and lager. "There's no need to get terribly complex."

BLUE CHEESES match well with any wine with residual sweetness — port, sherry, sauternes. The combination is always a crowd-pleaser. For beers, go with big, barrel-aged imperial stouts or barley wines. "You want something with some big booze." For cider, choose a funkier one with barnyard flavors. Or go with really sweet cider like an iced cider.

AGED CHEESES can handle more complex wine, so try them with bigger reds like Bordeaux, Cabernet-Merlot blends, Côtes du Rhônes, Rioja. Ciders and malty beers are other excellent options — think Belgian dubbel or a German Doppelbock, medium weight in alcohol but richer malty notes. If you like IPAs, "the red wines of the beer world," the hoppy overtones amplify any flavors in the cheese, so if there's a bitter note, the hops will make it extra-bitter. An IPA can work with a sharp cheddar or a smoked blue, but it's a tricky match.

SHEEP'S-MILK CHEESE, like aged cheeses, works best with bigger reds and ciders, but not so much beer.

STINKY WASHED-RIND CHEESES beg for aromatic wines like Riesling and Gewürztraminer and also ciders for funk on funk. Beer goes very nicely with these cheeses. A saison is a top pairing, but malty choices like dubbel are also good options.

12

BREAKFAST, BRUNCH, AND BREADS

RICOTTA
HOTCAKES

Ricotta Hotcakes
with Lemon Curd

These hotcakes are among that special breed of guestworthy brunch dishes that are also easy enough to prepare on a busy weekday. The cakes will be their fluffiest when made with ricotta that's fairly dry, so you may want to drain the cheese in a cheesecloth-lined sieve for several hours (or overnight) before you begin.

The pancakes also straddle sweet and savory in delightful fashion, so we've included lemon curd for the former — with or without fresh berries on top — and serving suggestions, below, for the latter.

MAKES 1 DOZEN

⅓ cup all-purpose flour

¾ teaspoon baking powder

Pinch of kosher salt

2 egg yolks plus 1 egg white

½ cup whole milk

Zest of 1 lemon

Zest of 1 orange

1 cup ricotta

Butter or oil, for griddle

1. Whisk the flour, baking powder, and salt together in a medium bowl. Mix in the egg yolks, milk, and both zests until smooth. Gently stir in the ricotta.

2. Whisk the egg white (by hand or with an electric mixer) in another bowl until stiff peaks form. Using a flexible spatula, fold the white gently but thoroughly into the ricotta mixture.

3. Heat a greased griddle or large cast-iron skillet until a drop of water bounces on its surface. Drop ¼ cup batter per cake onto griddle. Cook about 2 minutes, or until the bottoms of the cakes are golden brown, then flip and cook the other side until browned, about 2 minutes longer.

4. Serve immediately or place on a baking sheet fitted with a wire rack and keep warm in a 200°F (95°C) oven.

Savory Variation

Omit orange zest (and lemon zest, if desired) from the batter and add ½ teaspoon freshly ground pepper. Top with wilted spinach or kale, poached eggs (optional), and grated or shaved Parmesan or Romano.

Cinnamon Rolls
with Crème-Fraîche Glaze

From-scratch cinnamon rolls are, admittedly, not an everyday option, but they will most decidedly make any day start off on a special — and wonderfully scent-filled — note. This recipe ups the scrumptious factor by topping the still-warm rolls with a singular glaze, made not with the usual milk but crème fraîche, for richness.

An overnight rise in the refrigerator boosts the flavor of the rolls, but for more immediate cinnamon-roll satisfaction, proof the rolls on the counter for about 2 hours, or until doubled in bulk.

Recipe continues on next page

Cinnamon Rolls, *continued*

MAKES 1 DOZEN

FOR THE ROLLS

　　1 cup whole milk

　　2 eggs

　　½ cup granulated sugar

　　2 packages (¼ ounce each) or 2 tablespoons active dry yeast

　1½ teaspoons kosher salt

　3¾ cups all-purpose flour, plus more for dusting

　　½ cup whole-wheat flour

　　1 cup (2 sticks) unsalted butter, softened, plus more for baking pan

　　1 cup packed light brown sugar

　　3 tablespoons ground cinnamon

FOR THE GLAZE

　　½ cup crème fraîche, more if needed

　　1 cup confectioners' sugar

　　　Pinch of kosher salt

1. Make the rolls: Beat the milk, eggs, granulated sugar, yeast, and salt in the bowl of a stand mixer with the paddle attachment on medium speed until combined. Beat in one-third of the all-purpose flour.

2. Switch to the dough hook. Add the remaining all-purpose flour and the wheat flour. Mix on medium-low speed until the dough comes together. Increase speed to medium-high and add half the butter, a few pieces at a time, and mix until fully incorporated.

3. Cover bowl and let the dough rise in a warm, draft-free place until doubled, 1½–2 hours.

4. Put the remaining butter, the brown sugar, and cinnamon in a medium bowl and stir with a flexible spatula until combined.

5. Transfer dough to a lightly floured work surface. Using your hands and a rolling pin, gently press and roll the dough into an 18-by-24-inch rectangle. Spread the butter mixture across the dough, leaving a 1-inch border around the edge. Roll the dough tightly up into an 18-inch-long log. Cut the log into 12 (1½-inch-thick) slices.

6. Grease a 9-by-13-inch baking pan. Nestle the rolls into the pan and cover the pan with plastic wrap. Let rolls rise in the refrigerator overnight until doubled. Remove from refrigerator and let rolls sit at room temperature for about 30 minutes.

7. While the rolls are resting, preheat the oven to 350°F (180°C). Bake the rolls for about 30 minutes, or until golden brown and cooked through. (If the rolls start to brown too quickly, cover loosely with foil and reduce heat to 325°F/170°C.) Remove from oven and let cool while making the glaze.

8. Make the glaze: Whisk the crème fraîche, confectioners' sugar, and salt in a small bowl until smooth and combined. Add more crème fraîche as needed to reach the desired consistency. Spread over the warm rolls and serve.

CINNAMON ROLLS

Migas

Migas, which means "breadcrumbs" in Spanish, is a breakfast classic — especially in Austin, Texas, where the Tex-Mex version of the dish is believed to have been created. As with any respectable recipe that's built on leftovers, it's a natural for improvising with whatever happens to be in your larder (doubly so for the toppings). For the essential "crumbs," soft corn tortillas are the most traditional and yield a tender dish, while tortilla chips (either homemade or from a bag) will result in migas with a more crispy bite.

This version calls for two types of cheese, grated Monterey Jack and queso fresco, but you could go with one or the other, or swap in whatever Spanish-style cheese you have on hand.

SERVES 4

4 tablespoons corn or vegetable oil

1 white onion, finely diced

2 red, yellow, or orange bell peppers (or a combination), ribs and seeds removed, finely diced

1 cup chopped corn tortillas (about 4 tortillas) or crumbled tortilla chips

8 eggs, lightly beaten

Kosher salt and freshly ground pepper

1½ cups grated Monterey Jack or cheddar

FOR SERVING

½ cup salsa

1 ripe but firm avocado, peeled, pitted, and sliced

1 cup (about 2 ounces) crumbled queso fresco

½ cup chopped cilantro

6 scallions, thinly sliced

1 lime, cut into wedges

1. Heat a large cast-iron skillet over medium-high heat. Add the oil and when very hot, add the onion and bell peppers. Sauté, stirring occasionally, until onion is tender and starting to brown, about 4 minutes. Add the tortillas and stir to combine.

2. Pour the eggs into the skillet, season with salt and pepper, and cook, stirring and softly scrambling them with a spatula, until nearly cooked but still a little wet. Remove from heat and stir in the grated cheese.

3. Divide the migas among four plates and top each with salsa, avocado, queso fresco, cilantro, scallion, and a lime wedge. Serve immediately.

Pão de Queijo
(Brazilian Cheese Bread)

These cheesy puffs are delicious on their own, split and filled with cheese and cured ham, or as a tasty nibble alongside a bowl of soup. Formed into tiny rounds, the puffs are a gluten-free alternative to Gougères (page 311).

MAKES 2 DOZEN LARGE OR 4 DOZEN SMALL PUFFS

1 cup whole milk

½ cup corn or vegetable oil

1 teaspoon kosher salt

2 cups tapioca flour

2 eggs

1½ cups (about 6 ounces) grated Parmesan or Manchego

1. Preheat the oven to 450°F (230°C). Bring the milk, oil, and salt to a boil in a medium saucepan

over medium heat. Remove from the heat as soon as large bubbles start to form. Add the tapioca flour and stir in with a wooden spoon until there are no dry patches of flour. The dough will be grainy.

2. Transfer the dough to the bowl of a stand mixer fitted with the paddle attachment and beat the dough on low speed until it cools, about 2 minutes — it should be warm, but not too hot to touch. Beat the eggs, one at a time, into the dough on medium speed until they are incorporated. Beat in the cheese.

3. Line two baking sheets with parchment. Scoop tablespoons (for small puffs) or scant ¼ cups (for large puffs) of dough onto the sheets, leaving 1 inch between small puffs, 2 inches for large.

4. Bake for 20–30 minutes, depending on the size of the puff. The puffs will be crisp on the surface and slightly moist inside. Serve warm.

Acharuli Khachapuri
(Georgian Cheese Bread)

It's a wonder these boat-shaped flatbreads haven't hit the world food scene, given just how delectable they are. Sulgani is the traditional cheese in Georgia, and by all means you can use that if you've made some yourself. This modified version approximates that distinctive cheese by combining a melting cheese such as mozzarella with a fresher, tangier, crumbly cheese like feta. Great on their own, the breads are even better — and more authentic — when topped with an egg.

SERVES 4

⅔ cup warm water (100–115°F/38–46°C)

1 teaspoon active dry yeast (from 1 package)

1 teaspoon honey

1 tablespoon extra-virgin olive oil, plus more for bowl

1¼ cups all-purpose flour, plus more for dusting

1 teaspoon kosher salt

2 cups (about 8 ounces) shredded mozzarella or taleggio

1 cup (about 4 ounces) crumbled feta or ricotta salata

4 eggs (optional)

4 tablespoons unsalted butter, cut into cubes

1. Combine the warm water, yeast, honey, and oil in a large bowl. Let stand until foamy, about 5 minutes. Add the flour and salt and stir to make a soft dough. Turn the dough out onto a lightly floured work surface and knead until smooth and elastic. Transfer to a lightly greased bowl, turn to slick with oil, cover the bowl with plastic wrap, and let rise in a warm, draft-free place until dough is doubled, 45–60 minutes.

2. Preheat the oven to 500°F (260°C), with a pizza stone, baking steel, or heavy baking sheet in the bottom third of the oven. Mix the cheeses together.

3. Turn the dough out onto a lightly floured work surface. Cut dough into four equal pieces and set three aside. Roll the first piece of dough into a ⅛-inch-thick round. Place one-quarter of the

Recipe continues on next page

cheese in the center of round, spreading a little cheese slightly from the mound in the center and leaving a ½-inch border around the edge of the dough. Starting on one side of the round, roll the edge of the dough evenly toward the center, stopping about an inch before you reach the middle. Repeat on the other edge, leaving about 2 inches of cheese-covered dough exposed. Pinch the open ends of the rolls together, forming an oval with sharply tapered ends. Transfer to a parchment-lined baking sheet or a floured pizza peel. Repeat with the remaining dough and cheese.

4. Transfer the loaves to the pizza stone, spacing them out evenly. Bake for about 12 minutes, or until just golden brown. Remove from the oven, crack an egg into each parcel if desired, and return to oven until egg is set, about 4 minutes. Remove from oven, dot with butter, and serve.

CREAM CHEESE MUFFINS

These rich and delicious muffins are best the day you bake them, but will keep, wrapped well, at room temperature for another day or two. Serve slathered with — what else? — more cream cheese, and drizzle with honey if you can handle all that scrumptiousness.

MAKES 1 DOZEN

- 4 tablespoons unsalted butter, melted, plus more for pan
- 2 cups sifted all-purpose flour
- ½ cup sugar
- 3 teaspoons baking powder
- ½ teaspoon kosher salt
- ¾ cup (6 ounces) cream cheese, room temperature
- 2 eggs
- ⅔ cup milk
- 1 teaspoon vanilla extract
- ½ cup coarsely chopped pitted dates

1. Preheat the oven to 425°F (220°C). Grease 12 cups of a standard muffin tin. Whisk together the flour, sugar, baking powder, and salt in a large bowl.

2. Beat the cream cheese and eggs in a medium bowl until combined, then gradually beat in the milk and butter until smooth. Stir in the vanilla.

3. Make a well in the flour mixture and pour in the cream cheese mixture. Blend with a wooden spoon until just moistened. The batter should be lumpy. Stir in the dates.

4. Divide the batter evenly among the muffin cups and bake for 20–25 minutes, or until a tester inserted near the centers comes out clean. Let cool on a wire rack slightly before removing from the tin. Serve warm or at room temperature.

Italian Feather Bread
with Whey

This bread is always a big hit at cheese-making workshops. It also solves the constant cheese-making dilemma of what to do with all that whey!

MAKES 2 LOAVES

- 1 cup warm water (100–115°F/38–46°C)
- 2 packages (¼ ounce each) or 2 tablespoons active dry yeast
- 1 tablespoon sugar
- 5 tablespoons unsalted butter, cut into small pieces, plus more for pans
- ¾ cup hot whey or milk (120°F/49°C)
- 2 teaspoons kosher salt
- 5½–6 cups sifted all-purpose flour, plus more for dusting
- Cornmeal, for sprinkling
- 1 egg white, lightly beaten

1. Combine the water, yeast, and sugar in a large mixing bowl. Let stand until foamy, about 5 minutes.

2. Meanwhile, melt the butter in the hot whey and let cool to lukewarm. Add the salt, then pour into the yeast mixture and stir to combine.

3. Stirring vigorously with a wooden spoon, add the flour, 1 cup at a time, until the dough starts to come away from the sides of the bowl. The dough will be soft and sticky.

4. Turn the dough out onto a lightly floured work surface. Using a bench scraper or a large metal spatula, scrape under the flour and the dough, fold the dough over, and press it with your free hand. Continue doing this until the dough has absorbed enough flour from the surface and is easy to handle.

5. Knead for 2–4 minutes, or until dough is very smooth, being sure to keep your hands well floured, as the dough will still be sticky. Let rest for 5 minutes.

6. Divide dough in half. Roll out each half into a rectangle, about 12 by 8 inches. Starting from a long side, roll up the rectangle tightly into a log, pinching the seams as you go.

7. Grease two baking sheets and sprinkle with cornmeal. Place a loaf on each sheet and let rise in a warm, draft-free place until doubled, 50–60 minutes.

8. Preheat the oven to 425°F (220°C). Brush the loaves with egg white and bake for 30–40 minutes, or until the loaves are golden on top and make a hollow sound when you tap the top with your knuckles. Let cool on the pan on a wire rack before serving.

GLOSSARY OF CHEESES

MELTING CHEESES

Emmental

Fontina

Gouda (younger versions)

Gruyère

Havarti

Monterey Jack

Mozzarella

Muenster

Provolone

Reblochon-style

Taleggio

CRUMBLING CHEESES

Asiago (younger versions)

Camblu

Cheddar (aged versions)

Chèvre (drier version)

Feta

Gloucester

Gorgonzola

Gouda (aged versions)

Stilton-style cheese

Swiss (aged versions)

GRATING OR SHAVING CHEESES

Asiago (aged versions)

Caciocavallo (aged version)

Dry Jack

Gouda (aged versions)

Grana Padano

Manchego

Montasio (aged versions)

Parmesan

Ricotta Salata

Romano

13

DIPS, SPREADS, AND SMALL BITES

MANCHEGO-STUFFED
PIQUILLO PEPPERS, *page 314*

Savory Fromage Blanc Dips

Fromage blanc is one of the easiest cheeses to make; see the recipe on page 63. Once you produce all you can eat, you will want to figure out new and innovative ways to use it, and here are some effortless ways to do just that.

Each dip or spread follows the same formula: Mix fromage blanc in a bowl with some flavorings, and allow it to sit at room temperature for an hour or two for the most flavor. The dips are best the same day, but will keep for up to two days in the refrigerator.

All are delectable, but the blue cheese dip is my favorite, and I like to serve it in a beautiful dish surrounded by fresh carrots, celery, radishes, and tomatoes, and garnished with sprigs of parsley. You can use most any type of blue cheese to make it.

EACH MAKES ABOUT 1 CUP

Blue Cheese

Combine 1 cup (8 ounces) fromage blanc, ¼ cup (1 ounce) crumbled blue cheese, and 1 tablespoon snipped fresh chives in a bowl and blend well. Cover and refrigerate for up to 3 days. Whisk before serving.

Basil

Combine 1 cup (8 ounces) fromage blanc, 3 tablespoons finely chopped fresh basil leaves, ¼ teaspoon kosher salt, and a pinch of paprika in a bowl and blend well. Cover and refrigerate for up to 3 days. Whisk before serving.

Garlic

Combine 1 cup (8 ounces) fromage blanc, 2 tablespoons snipped fresh chives, 1 crushed garlic clove, and ¼ teaspoon kosher salt in a bowl and blend well. Cover and refrigerate for up to 3 days. Whisk before serving.

Fresh Dill

Combine 1 cup (8 ounces) fromage blanc, 2 tablespoons fresh dill leaves, 1 finely chopped scallion (white and green parts), and ¼ teaspoon kosher salt in a bowl and blend well. Cover and refrigerate for up to 3 days. Whisk before serving.

Fromage Fort

As a delicious way to use up little nibbles of cheese — and the last dregs of wine — from a party, fromage fort (French for "strong cheese") is a frugal home cheese maker's dream come true. It is utterly delicious spread on sandwiches, broiled atop crusty bread, or stuffed into endive or radicchio leaves. When making fromage fort, you can (honestly) use any and all types of cheese and in any combination.

MAKES ABOUT 1 CUP

8 ounces cheese, grated, crumbled, or cut into small pieces (about 2 cups)

1 garlic clove

Pinch of freshly ground pepper

2–4 tablespoons dry white wine

Kosher salt (optional)

Pulse the cheese, garlic, and pepper in a food processor to break up the cheese, then pour in 2 tablespoons of the wine with the processor running. Add more wine if the cheese seems too dry: It should be creamy and spreadable. Taste the cheese and add salt if needed (this is unlikely as the cheese will be pretty salty on its own). Pack into small jars or serving crocks and cover tightly. Store in the refrigerator for up to 1 month.

KOPANISTI
(GREEK RED-PEPPER SPREAD)

In Greece, kopanisti is both a fermented cheese with a very piquant, almost spicy flavor, and a savory spread that's meant to replicate that hard-to-find-cheese's qualities. This version of the latter is a natural addition to a mezze platter, alongside toasted pita, sliced cucumbers, and mixed olives.

MAKES ABOUT 2 CUPS

- 8 ounces feta, preferably from sheep's milk
- ¼ cup chopped roasted red pepper
- ¼ cup chopped pepperoncini
- ¼ cup extra-virgin olive oil
- 1 garlic clove
- ½ teaspoon dried mint, preferably Greek
- ½ teaspoon dried oregano, preferably Greek
- ¼–1 teaspoon red pepper flakes

Pulse feta, roasted pepper, pepperoncini, oil, garlic, mint, oregano, and pepper flakes in a food processor to combine. Pack into jars or serving crocks and cover tightly. Store in the refrigerator for up to 2 weeks.

PIMENTÓN CHEESE

There's the ubiquitous all-American pimento cheese spread that comes in a jar and is made with yellow cheddar (or cheese product) and those little strips of namesake red pepper. And then there's this Spanish-inspired version that uses a mix of white cheddar and Manchego for the cheese, and piquillo peppers instead of the usual ones, plus pimentón for flavor — and the play on words.

MAKES ABOUT 2 CUPS

- 4 ounces extra-sharp cheddar, grated (about 1 cup)
- 4 ounces Manchego, grated (about 1 cup)
- ¼ cup mayonnaise
- 1 (4-ounce) jar piquillo peppers, drained and finely chopped
- ¼ cup finely sliced scallions (white and green parts)
- 1 garlic clove, minced or pressed
- ½ teaspoon sweet or hot pimentón
- Kosher salt and freshly ground pepper

Combine both cheeses and the mayonnaise in a medium bowl. Stir until well combined. Add the piquillo peppers, scallions, garlic, and pimentón. Season with salt and pepper. Cover and store in the refrigerator for up to 2 weeks.

Burrata

with Sautéed Peppers and Bagna Cauda

Burrata, a version of mozzarella that oozes with extra cream, is luscious enough with just crusty bread, but why stop there? Red peppers sautéed in a classic bagna cauda take it to a whole other level.

SERVES 4

- 5 tablespoons extra-virgin olive oil
- 2 tablespoons unsalted butter
- 4 garlic cloves, thinly sliced
- 8 anchovy fillets
- 3 red bell peppers, ribs and seeds removed, cut into fine strips
- 8 ounces burrata
- ¼ cup finely chopped fresh flat-leaf parsley
- Grilled bread, for serving

1. Combine 2 tablespoons of the oil with the butter, garlic, and anchovies in a small saucepan. Cook over medium-low heat, stirring frequently, until anchovies start to break up, 2–3 minutes.

2. Heat the remaining 3 tablespoons oil in a large skillet over medium-high heat. When the oil is hot but not smoking, add the peppers. Stir and toss the peppers occasionally until starting to blister, soften, and brown, about 5 minutes.

3. Reduce the heat to medium-low. Nestle the burrata into the peppers and cover the pan. Cook for 5 minutes. Uncover the pan; drizzle the burrata with the anchovy mixture and sprinkle with parsley. Serve family-style in the skillet with grilled bread on the side.

Frico

Think of these cheese crisps as (gluten-free) bases for all manner of toppings: chopped fresh tomatoes with basil, roasted eggplant with thyme, or sautéed kale and garlic, to name just a few. Basically, anything you'd put on crostini or toasts works fine with frico, too — minus the cheese.

Frico are traditionally formed into cupped shapes (similar to tuile cookies) when cooling, but you can also just serve them as flat disks. Either way, frico are the epitome of cheesy goodness.

MAKES ABOUT 6

- 4 ounces Parmesan, Montasio, Grana Padano, or other hard grating cheese, grated (about 1 cup)
- ½ teaspoon caraway, cumin, fennel, or mustard seeds (optional)

1. Preheat the oven to 375°F (190°C). Line two baking sheets with a nonstick baking mat or parchment paper. Have ready a rolling pin or wine bottle (for shaping frico).

2. Mix cheese with spices, if desired. Sprinkle 2 tablespoons of cheese onto the baking sheet, forming a 4-inch round, spreading evenly. Repeat with the rest of the cheese, spacing rounds at least 1 inch apart.

3. Bake one sheet at a time until the cheese melts and just begins to turn golden. Immediately transfer frico to the rolling pin to cool and set the shape. Serve immediately or store in an airtight container at room temperature for 2–3 days.

GOUGÈRES

Not your everyday cheese puff: These crispy, fluffy, scrumptious bites are most often served with cocktails, though they are equally good as an accompaniment to soup or as a way to add heft to main-course salads (in lieu of croutons, for example). Gruyère is the classic choice for the cheese, but really any melting cheese — try fontina or Emmental — are just as worthy. You can also add flavorings as suggested below.

Once scooped onto baking sheets, gougères can be frozen until firm and then transferred to resealable plastic bags; they'll keep for three (or more) months. Bake straight from the freezer, for about five minutes longer than directed.

MAKES ABOUT 1 DOZEN

¼ cup water

¼ cup milk

4 tablespoons unsalted butter, cut into small pieces

Kosher salt and freshly ground pepper

½ cup all-purpose flour

2 eggs

2 ounces plus 2 tablespoons Gruyère, grated (about ¼ cup)

1. Preheat the oven to 400°F (200°C). Combine the water, milk, butter, and a generous pinch each of salt and pepper in a medium saucepan, and bring to a boil. Remove from heat.

2. Add the flour and stir quickly with a wooden spoon to combine. Put the pan back on the heat and stir until the dough forms a ball that pulls away from the sides and bottom of the pan.

3. Transfer the dough to the bowl of a stand mixer fitted with the paddle attachment and let cool for 5 minutes. With mixer on medium speed, beat in the eggs one at a time. The dough may break apart at some point in the mixing, but it will come together again with more beating.

4. Stir all but 2 tablespoons of the cheese into the dough to evenly distribute. Scoop (or pipe) heaping tablespoons of dough onto a parchment-lined baking sheet. Sprinkle the remaining cheese over the tops.

5. Bake for 20–30 minutes, until gougères are puffed and deeply golden. Serve warm.

FLAVOR VARIATIONS

- *Mix ¼ cup of finely chopped fresh herbs, such as thyme, rosemary, parsley, or oregano (or use a mixture), into the dough after adding the cheese.*

- *Substitute cheddar for Gruyère. Whisk 2 tablespoons cornmeal and a generous pinch of chili powder into the flour before proceeding with step 2.*

- *Stir 2 tablespoons finely chopped sun-dried tomatoes and 1 teaspoon minced fresh oregano into the dough in step 4, along with the cheese.*

- *Replace Gruyère with Manchego and season flour with 2 teaspoons pimentón.*

Beet Tartlets
with Ricotta and Blue Cheese

Beets and blue cheese are mutual-appreciation fans; here the two are paired in a tasty topping for savory tarts, with ricotta cheese adding extra creaminess. These tartlets start in a muffin tin to create single servings that can be served alongside a simple salad as a knife-and-fork starter course. Or, for bite-sized finger foods, bake the tartlets in mini muffin tins.

A mix of beets — golden, red, and Chioggia (the gloriously stripey ones) — will make the most dramatic presentation, but you can use whatever kind of beets you can find. It's fine to roast the beets a day or two before assembling the tarts, too, but be sure to peel them while still warm.

MAKES 12

Nonstick cooking spray

Pâte Brisée (recipe follows)

Flour, for dusting

12 medium beets (about 3½ pounds), trimmed and scrubbed

¼ cup extra-virgin olive oil, plus more for drizzling

Kosher salt and freshly ground pepper

8 ounces blue cheese (any kind)

¾ cup (6 ounces) ricotta, well drained

½ cup grated Parmesan (about 2 ounces)

1. Coat a standard muffin tin with cooking spray. Remove one dough disk from refrigerator. Dust a work surface lightly with flour and roll the dough out to a ⅛-inch thickness.

2. Using a cookie cutter or a sharp knife, cut six 5-inch rounds of dough. Line the muffin-tin cups with the rounds, crimping the top edge of the dough. Repeat with the second disk of dough. Pierce each round all over with the tines of a fork. Freeze for 30 minutes.

3. Meanwhile, preheat the oven to 375°F (190°C). Bake until crust is golden brown, about 25 minutes. Let cool in tins on a wire rack. Leave the oven on.

4. Toss beets with oil and sprinkle with salt and pepper. Spread evenly in a roasting pan or rimmed baking sheet and cover with foil. Roast for about 1 hour, or until tender enough to be pierced with a paring knife. Remove from oven and let cool slightly before rubbing off skins using paper towels (use a paring knife for tough spots). Cut beets in half and then slice into ⅛-inch-thick half-moons using a mandoline or very sharp knife.

5. Mix the blue cheese and ricotta together, then season with a generous pinch of pepper. Divide cheese mixture evenly among the tartlet shells.

6. Divide beet slices among tartlets, arranging them in an overlapping fashion to cover (some beets may be left over).

7. Sprinkle with Parmesan and drizzle with oil. Bake for 30 minutes, or until crust is deep golden. Let cool slightly before serving.

Pâte Brisée

4½ cups all-purpose flour, plus more for dusting

¾ teaspoon kosher salt

½ teaspoon sugar

1½ cups (3 sticks) unsalted butter, cut into cubes, well chilled

¾ cup cold water or whey, plus more as needed

1. Pulse flour, salt, sugar, and butter together in a food processor until mixture looks like coarse crumbs with only pea-sized pieces of butter, 15–20 pulses.

2. With the processor running, pour in nearly all of the water, processing until the dough forms a ball. Add additional water if needed.

3. Turn dough out onto a lightly floured work surface and dust with flour. Gather into a ball. Divide in half and press each into a disk. Dust with flour and wrap in plastic. Chill for at least 1 hour or overnight. Dough can also be frozen for up to 3 months; thaw in refrigerator.

Feta and Mint Borek

Borek is one of many dishes in the family of phyllo-wrapped Middle Eastern pastries. For this recipe, sheets of yufka — precut triangles, squares, or rounds of thin pastry — are used instead of the more familiar phyllo. If you cannot find yufka, use horiatiki ("country-style") phyllo; its thicker layers hold up better than standard phyllo in this dish.

MAKES 16 SMALL HORS D'OEUVRES OR 8 APPETIZER-SIZED PASTRIES

2 cups crumbled feta (about 8 ounces)

¼ cup chopped fresh mint

½ cup chopped fresh flat-leaf parsley

1 teaspoon dried mint, preferably Greek

Kosher salt (optional) and freshly ground pepper

1 (12-ounce) package triangular yufka sheets or country-style phyllo

½ cup (1 stick) unsalted butter, melted, or ½ cup extra-virgin olive oil

Vegetable oil, for frying

1. Mix the cheese and herbs in a bowl, then season with salt, if needed, and several grinds of pepper.

2. Place one sheet of yufka on work surface, keeping remaining sheets covered with plastic and damp towel. Brush sheet with butter. If using phyllo, unroll and cut it in half the long and short ways to make four large rectangles.

3. Spoon 1 tablespoon filling evenly onto wide end of dough triangle, leaving a 1-inch margin. Fold sides in over filling. Starting at filling end, roll toward the pointed end, forming a cigar shape. Place on baking sheet. Repeat with remaining filling and pastry and brush with butter.

4. Pour vegetable oil into large skillet to a depth of ½ inch and heat to 350°F (180°C). Working in batches, fry pastries until golden brown, about 2 minutes per side. Drain on paper towels before serving.

Fondue for Friends

Fondue may have had its moment in the 1970s, but this melting-pot dish will never truly fall out of favor, and for good reason: It's always a crowd-pleaser, especially during cold-weather months and in front of a roaring fire. If you happen to have a fondue pot in the pantry, dust it off and use it for serving, though a slow cooker, chafing dish, or attractive ceramic casserole set over a warming candle can also do the job quite nicely.

Fondue making is a fairly simple process and is infinitely adaptable. Here are two options, and the method is the same for both. Be sure to offer plenty of toasted bread cubes, apple or pear slices, steamed potatoes, and/or cut vegetables, for dredging and delighting.

SERVES 6

FOR TRADITIONAL SWISS GRUYÈRE FONDUE

12 ounces Gruyère, grated (about 3 cups)

12 ounces Emmental, grated (about 3 cups)

1½ tablespoons all-purpose flour or tapioca flour (for a gluten-free option)

1½ cups dry white wine

1 tablespoon kirsch (optional; or substitute apple cider)

Pinch of freshly grated nutmeg

Pinch of freshly ground pepper

FOR A NEW CLASSIC GOUDA-BEER FONDUE

1½ pounds Gouda, grated (about 4½ cups)

1½ tablespoons all-purpose flour

1½ cups amber ale

1 tablespoon bourbon (optional; or substitute apple cider)

Pinch of freshly ground pepper

1. Toss the cheeses and flour together in a large bowl.

2. Pour the white wine or beer, optional spirits, nutmeg, and pepper into a medium saucepan and bring to a simmer over low heat. Gently whisk in the cheese mixture in handfuls until smooth and melted. Do not let the cheese boil.

3. Transfer cheese mixture to a fondue pot or other serving vessel and serve with desired dippers.

Manchego–Stuffed Piquillo Peppers

With all due respect to jalapeño peppers, these updated stuffed peppers just might become your new favorite cocktail-time treat. What's more, piquillo peppers (from the jar) are already roasted, peeled, and seeded, for an effortless appetizer — not counting time spent making the Manchego, that is.

MAKES 6

4 ounces Manchego, cut into ½-inch cubes (about 1 cup)

¼ cup quince jam

1 (9- to 12-ounce) jar piquillo peppers, drained

2 tablespoons extra-virgin olive oil

½ cup sliced almonds

8–12 small slices of toasted rustic bread (optional)

1. Mix the cheese and jam together in a small bowl. Carefully stuff each pepper with the cheese mixture, dividing it equally between the peppers.

2. Heat the oil in a medium cast-iron skillet over medium-high heat. Pat the peppers dry with paper towels and lay them gently in the skillet. Scatter the almonds around the peppers. Cook the peppers on one side for about 4 minutes, then carefully flip and cook another 3 minutes.

3. Serve the peppers topped with the almonds, either alone or on toasted bread for a tapas-style presentation.

SOUTHERN-STYLE CHEESE STRAWS

Traditionalists might demand you use cheddar — preferably the kind with annatto added — but there are so many other aged cheeses to choose from. Gruyère, Manchego, and Asiago are all excellent options, but really you should make them with whatever you have in your aging room. These nifty nibbles just beg to be served with a Bloody Mary or other brunch cocktail of choice.

MAKES 18

1½ cups all-purpose flour, plus more for dusting

1½ teaspoons dry mustard

1½ teaspoons cayenne pepper

1 teaspoon kosher salt

8 ounces cheddar cheese, grated (about 2½ cups)

½ cup (1 stick) unsalted butter, softened

1. Whisk flour, mustard, cayenne, and salt together in a medium bowl.

2. Blend cheese and butter with an electric mixer on medium speed until smooth and combined. With mixer on low speed, gradually beat in flour mixture until combined. (Alternatively, blend cheese and butter in a food processor, then add flour mixture in three batches, pulsing until combined after each addition.)

3. Turn dough out onto a lightly floured surface and roll out into a 12-by-9-inch rectangle. Chill on a parchment-lined baking sheet until firm, about 15 minutes.

4. Preheat the oven to 425°F (220°C), with racks in upper and lower thirds. Remove dough from refrigerator and cut the rectangle in thirds crosswise (for three 4-by-9-inch pieces). Using a pastry wheel, pizza cutter, or sharp knife (and a ruler or other straight edge as a guide, if desired), cut the dough lengthwise into strips, each about ½ inch wide.

5. Divide strips between two parchment-lined baking sheets, about 2 inches apart, and bake until crisp, 12–15 minutes, rotating the sheets from top to bottom rack halfway through. Let straws cool for 1 minute on sheet, then transfer straws to a wire rack to cool completely before serving. Store in an airtight container at room temperature for up to 1 week.

14

SOUPS, SALADS, AND SIDE DISHES

SPICED
PUMPKIN SOUP

SPICED PUMPKIN SOUP
WITH PANEER CROUTONS

Paneer's firm texture stands up well to pan-searing, becoming a delightful (and gluten-free) stand-in for a more typical "crouton." Here, the toothsome tidbits are used to garnish an Indian-inspired soup that's heady with warm spices and enriched with coconut milk. Be sure to use sugar pumpkins instead of the large carving pumpkins for Halloween; butternut or kabocha squash will also work here.

SERVES 8

- 4 tablespoons vegetable oil
- 1 large yellow onion, chopped
- ½ teaspoon ground cumin
- ½ teaspoon ground coriander
- ¼ teaspoon ground cinnamon
- ¼ teaspoon cayenne pepper
- Pinch of ground cloves
- 6 cups cubed peeled pumpkin (from about 3 pounds)
- 1 (15-ounce) can unsweetened coconut milk
- 4 ounces paneer, cut into ½-inch dice (about 1 cup), dried well on paper towels
- 1 teaspoon cumin seeds
- ¾ cup torn or chopped cilantro

1. Heat 2 tablespoons of the oil in a large saucepan or stockpot over medium-high heat. Sauté the onion until just starting to soften, about 8 minutes. Stir in the cumin, coriander, cinnamon, cayenne, and cloves, and continue to cook, stirring, until spices are fragrant, about 2 minutes.

2. Add the pumpkin and coconut milk to the pot along with just enough water to barely cover the pumpkin. Simmer until pumpkin is tender, about 20 minutes. Purée with an immersion blender until soup is very smooth. Alternatively, purée in batches in a regular blender, being careful not to fill jar more than halfway.

3. Heat the remaining 2 tablespoons oil in a large nonstick skillet over medium-high heat. Add the paneer cubes and cumin seeds. Cook, stirring gently, until the paneer starts to turn golden.

4. To serve, ladle the soup into bowls and top with paneer and cilantro.

WATERCRESS VICHYSSOISE
WITH CRÈME FRAÎCHE

In this update of the French classic, crème fraîche is puréed into the soup and dribbled on top. This recipe also provides yet another excellent way to use up leftover cheese-making whey.

SERVES 6

- 2 tablespoons unsalted butter
- 2 leeks, cut into rings and washed well (about 2 cups)
- 1½ cups diced peeled russet potatoes (about 1 large)
- 3 cups whey, water, or vegetable broth
- 1¼ cups crème fraîche
- 2 bunches watercress, stemmed and coarsely chopped
- Kosher salt and freshly ground pepper

Recipe continues on next page

1. Melt the butter in a stockpot over medium-high heat. Add the leeks and cook, stirring occasionally, until softened. Add the potato and whey, and bring to a simmer. Cook until potato is fork-tender, about 8 minutes.

2. Add 1 cup of the crème fraîche and the watercress to the pot and stir to combine and start wilting the watercress. Purée soup with an immersion blender. Alternatively, purée in batches in a regular blender, being careful not to fill jar more than halfway. Add salt and pepper to taste.

3. Chill soup, covered, at least 4 hours or preferably overnight. Before serving, season with more salt to taste, ladle into bowls, and top with remaining ¼ cup crème fraîche.

NOTE: For a velvety smooth consistency, strain the soup before puréeing in step 2; if the idea of wasting the solids leaves you cold, purée them with a bit of the strained soup and stir to combine.

SORREL SOUP (SCHAV)
WITH BUTTERMILK AND WHEY

Schav is the Yiddish name for sorrel as well as the traditional soup that features this obscure herb, which is widely available at farmers' markets and sometimes at more produce-friendly food stores. With its bright taste, sorrel is worth seeking out (or growing yourself), but if you can't find it, spinach is your best bet; squeeze lemon wedges over each serving for that welcome hit of sourness that the herb, and this soup, is renowned for.

Here, the soup is enriched with buttermilk, for even more tang (and creaminess, traditionally provided by sour cream), and thickened with potato and also whey — making this an especially good recipe for avid cheese makers.

SERVES 6

4 tablespoons unsalted butter

1 shallot, minced

2 cups diced peeled russet potatoes (about 2 medium)

2 cups whey (or water), plus more if needed

2 bunches of sorrel, cut into fine strips (about 4 cups)

2 cups buttermilk

Kosher salt and freshly ground pepper

Thinly sliced radishes and grated lemon zest, for serving (optional)

1. Melt the butter in a large saucepan or stockpot over medium-high heat. Sauté the shallot until softened, about 5 minutes. Add the potatoes and whey, and bring to a simmer. Cook until the potatoes are fork-tender, about 8 minutes.

2. Add the sorrel to the pot and stir until wilted. Stir in the buttermilk to combine. Season with salt and pepper. Transfer to a bowl and let cool completely.

3. Cover the bowl and chill soup overnight. Before serving, whisk the soup to loosen; if it is too thick, whisk in more whey (or water). Season with more salt to taste and serve topped with radishes and lemon zest, if desired.

Ribollita

with Parmesan Broth

Ribollita ("reboiled" in Italian) was created out of the desire to use up every last bit of leftover soups or stews. This version is prepared completely from scratch — and gets added flavor from the addition of leftover Parmesan rinds that simmer in the broth (if you don't have any of these remnants yet, now's a good time to start keeping them, wrapped tightly, in the freezer). That said, if there's a pot of minestrone in the refrigerator, follow the technique below to maintain the spirit of the original.

SERVES 4

- 2 tablespoons extra-virgin olive oil, plus more for drizzling
- 1 large yellow onion, chopped
- 1 medium carrot, chopped
- 3 garlic cloves, minced
- 4 cups chicken or vegetable broth
- 1–2 Parmesan rinds
- 1 (14-ounce) can chopped tomatoes with their juice
- 1 large sprig rosemary
- 1 bunch kale, stemmed and chopped into bite-sized pieces
- 2 cans (14 ounces each) cannellini beans, drained and rinsed
- Kosher salt and freshly ground pepper
- Red pepper flakes
- 4 slices rustic bread, toasted and torn into bite-sized pieces
- ¼ cup grated Parmesan

1. Heat the oil in a large saucepan or stockpot over medium-high heat. Add the onion, carrot, and garlic. Cook, stirring occasionally, until vegetables start to soften, about 8 minutes. Add the broth, Parmesan rind(s), tomatoes and juice, and rosemary, and simmer until vegetables are very tender, about 8 minutes more.

2. Stir in the kale and beans. Season with salt, a generous quantity of pepper, and pepper flakes to taste. Cook until kale is just wilted, about 3 minutes.

3. Stir the bread into the soup and simmer until most of the liquid is absorbed into the bread and the ribollita has a stewlike consistency. Season with more salt, as desired. Sprinkle with Parmesan, drizzle with olive oil, and serve (without rinds).

Fatoush

with Halloumi Croutons

The Middle Eastern answer to panzanella, fatoush employs the idea of using leftover bread — in this case pita — to make a satisfying main-course salad. As with any type of salad, it's only as good as the produce, which should be at the peak of freshness, especially the tomatoes. Otherwise, anything goes — meaning you can swap out the halloumi for feta, lemon juice for vinegar, grated lemon zest for sumac (the traditional Middle Eastern spice), and whatever vegetables you have on hand for the ones called for here. You can even use country bread in place of pita, in which case you're right back at panzanella.

Recipe continues on next page

Fatoush, *continued*

SERVES 4

2 pita breads, preferably day old

⅓ cup plus 3 tablespoons extra-virgin olive oil, plus more for pitas

2 medium tomatoes, cored and cut into chunks

1 large cucumber, halved lengthwise, seeded, and cut into thin half-moons

4 radishes, cut into thin rounds

4 scallions, thinly sliced

1 cup fresh flat-leaf parsley leaves

¼ cup fresh mint leaves, coarsely chopped

1 tablespoon sumac, plus more for sprinkling

1 small garlic clove, minced

Juice of 1 lemon (about 3 tablespoons)

Kosher salt and freshly ground pepper

6 ounces halloumi, cut into ½-inch cubes (about 1½ cups)

1. Preheat the oven to 400°F (200°C). Split the pita into four rounds. Rub with oil and bake until dry and starting to turn golden brown. Remove from oven and let cool.

2. Combine the tomatoes, cucumber, radishes, scallions, parsley, mint, sumac, and garlic in a large bowl. Break the toasted pita into the bowl and toss well to combine.

3. Drizzle lemon juice and ⅓ cup of the oil over the salad. Season with salt and pepper and toss well.

4. Heat the remaining 3 tablespoons oil in a large skillet. Press the halloumi between paper towels to remove excess moisture. When the oil is hot, carefully add the halloumi to the skillet in a single layer. Fry, stirring gently, until the cubes start to brown. Drain briefly on paper towels before gently stirring halloumi into the salad. Sprinkle the fatoush with additional sumac and serve.

KALE SALAD
WITH BUTTERNUT SQUASH AND MANCHEGO

In this kale salad, Spanish flavors rule the day, thanks to a smoky vinaigrette and sharp Manchego — not to mention plump dates and toasted pepitas. There will be more onion-pimentón dressing than needed for this salad; drizzle the leftovers over grilled fish, chicken, or vegetables.

SERVES 4

2 cups diced peeled butternut squash (from about 1 pound)

1 cup plus 2 tablespoons extra-virgin olive oil

Kosher salt

¼ cup sherry vinegar

2 teaspoons Dijon mustard

¼ cup minced red onion

2 teaspoons sweet or hot pimentón

4 cups loosely packed baby kale or shredded kale leaves (stemmed)

2 ounces Manchego, shaved

½ cup dates, pitted and chopped into small pieces

½ cup pepitas, toasted

Recipe continues on next page

KALE SALAD

Kale Salad, *continued*

1. Preheat the oven to 400°F (200°C). Toss the squash with 2 tablespoons of the oil and a pinch of salt. Spread the squash onto a baking sheet and roast for about 20 minutes, until golden brown and tender. Remove from oven. Reserve some for garnish and place the rest in a large bowl.

2. Purée the vinegar, mustard, onion, and pimentón in a blender until chunky. With the blender running, drizzle in the remaining 1 cup oil until the dressing emulsifies. Season with salt.

3. Add the kale and most of the cheese to the bowl with the squash. Toss with just enough dressing to coat the leaves. Serve garnished with the dates, pepitas, and remaining squash and cheese.

Farro Salad
with Crème-Fraîche Dressing

This make-ahead, better-the-next-day salad is adapted from one developed by Vermont Creamery. It showcases the bright tang of crème fraîche and crumbled goat cheese against fresh vegetables and earthy farro. If you don't have farro on hand, substitute barley, wheat berries, quinoa, or brown rice.

SERVES 4

1½ cups farro

Kosher salt and freshly ground pepper

1 bunch (about 1 pound) red or golden beets, peeled and cut into ½-inch-thick wedges

⅓ cup plus 1 tablespoon extra-virgin olive oil

2 tablespoons crème fraîche

Zest and juice of 1 lime

2 tablespoons fresh lemon juice

2 teaspoons chopped fresh thyme

1 teaspoon Dijon mustard

1 garlic clove, minced

2 cups frozen peas, thawed

4 radishes, thinly sliced

4 ounces goat cheese, crumbled (about 1 cup)

1. Preheat the oven to 400°F (200°C). Bring a large pot of water to a boil. Add the farro and a generous pinch of salt. Cook at a rolling boil (like pasta) until tender, 30–45 minutes. Drain well.

2. Meanwhile, on a rimmed baking sheet, toss the beets with 1 tablespoon of the oil and season with salt and pepper. Cover sheet tightly with foil and roast for about 30 minutes, until the beets are knife-tender. Remove foil and cook beets for another 10–15 minutes, until the pan is dry and the beets are lightly browned.

3. Combine the crème fraîche, lime zest and juice, lemon juice, thyme, mustard, garlic, salt and pepper to taste, and the remaining ⅓ cup oil in a medium jar. Cover jar and shake to combine. (If making ahead, refrigerate and shake again before using.)

4. In a large bowl, toss the cooked farro with half of the dressing. Add the peas along with half the radishes and goat cheese and toss again. Garnish with the remaining radishes and goat cheese, drizzle with remaining dressing, and serve.

Roasted Agrodolce Cauliflower Salad
with Goat Cheese

Agrodolce is the Italian word for the classic pairing of sweet and sour. Here, cauliflower is roasted until golden brown, then tossed with currants and caramelized onion, for sweetness, and lemon and vinegar, for sourness. Lentils and fresh goat cheese round out this hearty salad or side dish.

SERVES 8

1 cup lentils, preferably *lentilles du Puy*

1 head cauliflower, cut into bite-sized florets

½ cup plus 2 tablespoons extra-virgin olive oil

Kosher salt and freshly ground pepper

1 large yellow onion, diced

2 garlic cloves, minced

½ cup currants

Zest and juice of 2 lemons (about ¼ cup plus 2 tablespoons juice)

2 tablespoons red wine vinegar

½ cup fresh flat-leaf parsley leaves

4 ounces goat cheese, crumbled (about 1 cup)

1. Preheat the oven to 400°F (200°C). Bring a large pot of water to a boil. Add the lentils and cook, uncovered, until just tender (they should not be at all mushy), 30–45 minutes. Drain and set aside.

2. While the lentils are cooking, toss the cauliflower with 2 tablespoons of the oil. Spread cauliflower evenly on a rimmed baking sheet and season with salt. Roast cauliflower for about 15 minutes, until golden and tender.

3. Heat the remaining ⅛ cup oil in a large skillet over medium heat. Add onion and garlic, and cook, stirring occasionally, until onion is golden and soft, about 20 minutes. Transfer to a large bowl.

4. Add roasted cauliflower, lentils, currants, lemon zest and juice, and vinegar to the bowl. Toss in parsley leaves. Season with salt and pepper. Sprinkle the goat cheese over the salad, toss gently, and serve.

Tartiflette
(Potato, Bacon, and Cheese Gratin)

Savoy (or Savoie), France, is famous for its washed-rind cheeses, especially Reblochon — and also for hearty dishes that that use those cheeses to delicious effect. Tartiflette — tartifle means potato — is basically a potato gratin to the nth degree, given that it is larded with bacon and doused with white wine.

SERVES 6

2 tablespoons unsalted butter, plus more for baking dish

8 ounces slab bacon, cut into lardons

1 medium yellow onion, thinly sliced

½ cup dry white wine

2½ pounds waxy potatoes, peeled and thinly sliced

Kosher salt and freshly ground pepper

1 pound Reblochon-style cheese, thinly sliced

Recipe continues on next page

1. Preheat the oven to 375°F (190°C). Grease a 2- to 3-quart gratin or casserole dish with butter.

2. Heat the butter in a large skillet over medium heat until foamy. Cook bacon until starting to crisp, tossing occasionally, about 10 minutes. Using a slotted spoon, transfer bacon to paper towels to drain.

3. Add onion to skillet; cook, stirring frequently, until slightly caramelized, about 10 minutes. Add wine and bring to a boil. Cook until reduced by half, 2 or 3 minutes. Add potatoes and season with salt and pepper. Cook until potatoes are just tender, 8–10 minutes.

4. Spread half the potato mixture evenly in prepared dish; top evenly with half the bacon and half the cheese. Repeat with remaining potato mixture, bacon, and cheese; bake for about 25 minutes, until top is browned and filling is bubbling. Let cool slightly before serving.

Brussels Sprouts and Apple Salad

with Smoked Gouda

Fall flavors permeate this no-cook salad (well, if you don't count toasting the pecans). It would be at home any night of the week with roasted pork or chicken, or even on the Thanksgiving table with all the other trimmings. Smoked Gouda takes the dish to a whole other taste level, but regular Gouda (or even Edam or Emmental) is a more-than-acceptable stand-in.

Juice of 1 lemon (about 3 tablespoons)

1 tablespoon Dijon mustard

½ small shallot, minced

1 garlic clove, minced

Kosher salt and freshly ground pepper

¼ cup extra-virgin olive oil

1 pound Brussels sprouts, trimmed and shredded

2 ounces smoked Gouda, grated (about ½ cup)

½ Granny Smith apple, peeled, cored, and grated

¼ cup dried tart cherries, coarsely chopped

2 tablespoons pecans, toasted and finely chopped

1. Whisk lemon juice, mustard, shallot, and garlic together in a large bowl. Season with salt and pepper. Slowly pour in oil, whisking to combine.

2. Add Brussels sprouts and toss to coat. Let sit for at least 1 hour and up to 2 hours.

3. Add cheese, apple, and cherries, and toss gently to combine. Season with salt and pepper, sprinkle with pecans, and serve.

Buttermilk Salad Dressings

Homemade buttermilk is the tangy dividend when making butter — or according to the recipe on page 263. Fresh buttermilk is especially delicious as the creamy base for flavorful salad dressings.

The method is similar regardless of what dressing you choose: Start with fresh buttermilk, add a thickener of some sort, and season to taste. Combine all ingredients in a blender and purée until smooth, adding salt and pepper to taste.

Use these recipes as a starting point for your own creations.

EACH MAKES ABOUT 1½ CUPS

CREAMY BUTTERMILK RANCH

1 cup buttermilk

½ cup sour cream

2 tablespoons finely chopped fresh chives

2 tablespoons finely chopped fresh flat-leaf parsley

2 tablespoons finely chopped fresh thyme

1 teaspoon dry mustard

Kosher salt and freshly ground pepper

GREEN GODDESS DRESSING

1 cup buttermilk

½ cup sour cream

1 tablespoon fresh lemon juice

2 tablespoons finely chopped fresh tarragon

2 tablespoons finely chopped fresh flat-leaf parsley

1 or 2 anchovy fillets

Kosher salt and freshly ground pepper

AVOCADO BUTTERMILK DRESSING

1 cup buttermilk

½ ripe avocado, pitted, peeled, and cut into chunks

2 tablespoons finely chopped cilantro

¼ teaspoon ground cumin

Pinch of cayenne pepper

Kosher salt and freshly ground black pepper

CURRIED BUTTERMILK DRESSING

1 cup buttermilk

½ cup yogurt

2 tablespoons finely chopped cilantro

¼ teaspoon ground cumin

¼ teaspoon curry powder

Kosher salt and freshly ground pepper

15

MAIN COURSES

SKILLET PIZZA

SKILLET PIZZA

WITH GREENS AND PROSCIUTTO

Cooking a pizza in a skillet helps achieve the characteristic trait of any respectable pie: a crispy crust. Plus, it can go straight from the stovetop to the table, for family-style serving. You can vary the toppings based upon what's hiding in the refrigerator, but do try this combination — taleggio and prosciutto are a match made in pizza paradise, and the greens provide an interesting counterpoint. Feel free to swap store-bought dough for your own.

MAKES ONE 10-INCH PIZZA

- ½ pound Pizza Dough (recipe follows)
- All-purpose flour, for dusting
- 2 tablespoons extra-virgin olive oil, plus more for drizzling
- 2 garlic cloves, minced
- 1–2 teaspoons red pepper flakes, plus more for serving
- ½ bunch kale or Swiss chard, stemmed and chopped or cut into fine strips
- Kosher salt and freshly ground pepper
- 4 ounces taleggio, cut into ¼-inch-thick slices
- 2 ounces sliced prosciutto

1. Place the pizza dough on a lightly floured work surface and press or roll out into an 8-inch round. Let dough rest while preparing the greens.

2. Heat the oil in a 10-inch cast-iron skillet over medium-high heat. Cook garlic and pepper flakes, stirring constantly, until aromatic, about 30 seconds. Add kale, toss to coat, and sauté until wilted, 3–5 minutes. Season with salt and remove greens from pan.

3. Stretch the dough into an 11-inch round and nestle the dough evenly into the pan, folding over any excess. Place the dough-lined pan over medium heat. Arrange one-half of the cheese over the dough, top with greens, and then add the remaining cheese. Cover the pan with foil and cook until the cheese is melted and the crust is cooked through, about 10 minutes. Remove foil and cook for 2 minutes more, to dry out crust.

4. Arrange the prosciutto slices over the top, drizzle with more oil, and sprinkle with pepper flakes before serving.

PIZZA DOUGH

MAKES ENOUGH FOR TWO 10-INCH PIES

- 1 packet (¼ ounce) active dry yeast
- ¾ cup warm water (100–115°F/38–46°C)
- 2 tablespoons extra-virgin olive oil, plus more for bowl and brushing
- 1 tablespoon sugar
- 2 teaspoons kosher salt
- 2 cups all-purpose flour, plus more for dusting

1. Sprinkle yeast over the warm water in a large bowl and let stand until foamy, about 5 minutes.

2. Whisk oil, sugar, and salt into yeast mixture. Add flour and stir until a sticky dough forms. Transfer dough to an oiled bowl and brush top with oil. Cover with plastic wrap and let rise in a warm, draft-free place until doubled, about 1 hour.

3. Turn dough out onto a lightly floured work surface and gently knead 1 or 2 times before dividing in half.

Recipe continues on next page

Dough can be wrapped tightly and refrigerated overnight, or frozen in a resealable plastic bag for up to 3 months; let thaw overnight in refrigerator before using.

3-Cheese Macaroni
with Panko

It wouldn't be a cheese book without macaroni and cheese — the ultimate comfort food and cheese dish in one. Nonnegotiables include full-flavored cheeses, a crispy top, and a creamy sauce. Elbows are the classic pasta shape, but cavatappi have more grooves, all the better for holding more sauce.

A heavy cream reduction replaces the béchamel sauce, a technique borrowed from Ashley Christensen of Poole's Downtown Diner in Raleigh, North Carolina. Substitute other cheeses for those listed here; just make sure the total weight equals 1¼ pounds.

SERVES 4

Oil or butter, for baking dish

3 cups heavy cream

Kosher salt and freshly ground pepper

8 ounces cavatappi or elbow pasta

4 ounces Swiss or Emmental, grated (about 1 cup)

4 ounces Parmesan, grated (about 1 cup)

12 ounces sharp cheddar, grated (about 3 cups)

½ cup panko breadcrumbs

1. Preheat the broiler, with the rack 4–6 inches from the heat source. Grease a 3-quart casserole or baking dish.

2. Bring the cream to a simmer in a large saucepan over medium-high heat. Cook over medium heat until reduced to 2¼ cups, about 10 minutes. Keep an eye on the cream to ensure it doesn't boil over.

3. While the cream is reducing, bring a large pot of water to a boil. Add salt and cook the pasta according to package instructions a minute or two short of al dente. Drain the pasta and keep warm.

4. Mix the cheeses together in a medium bowl. Put one-quarter of the cheese mixture into another bowl and add the panko. Set the panko mixture aside for the topping.

5. When the cream has reduced, stir in the pasta. Add the first cheese mixture to the pan in handfuls, stirring after each addition to make sure it is evenly melted before adding more. After all the cheese mixture has been added, pour the contents of the pan into the prepared baking dish. Sprinkle with panko mixture.

6. Broil until the top is browned in spots, rotating the dish if necessary to ensure even browning and watching carefully to avoid scorching. Let cool slightly before serving.

Foolproof Cheese Soufflé

Cheese soufflés are too delectable to reserve just for entertaining; this streamlined version offers the same incredible flavor but without all the fuss. You can make it with any melting cheese (see page 306), but *Gruyère, Emmental, or cheddar are all excellent options — either on their own or, better yet, in a combination. Serve with a simple green salad or steamed asparagus.*

SERVES 4

6 tablespoons unsalted butter, plus more for baking dish

⅓ cup all-purpose flour

2 cups whole milk, room temperature

Kosher salt and freshly ground pepper

Pinch of freshly grated nutmeg

6 eggs

10 ounces melting cheese (see headnote), grated (about 2½ cups)

¼ cup finely chopped fresh chives or flat-leaf parsley

1. Preheat the oven to 400°F (200°C). Grease a 6-cup soufflé or gratin dish.

2. Melt the butter in a medium saucepan over medium heat. Whisk in the flour, making sure there are no lumps. Cook, stirring with a wooden spoon, for about 30 seconds. Pour in the milk, whisking constantly, until it boils. Remove from heat and season béchamel with salt, pepper, and nutmeg. Let cool a little.

3. Break the eggs into a large bowl. Whisk for about 1 minute to combine and aerate.

4. Stir the grated cheese into the béchamel in the pan, then add the eggs and chives and stir well to combine. Scrape mixture into prepared baking dish and bake for about 40 minutes, until puffed and golden. Serve immediately.

GNUDI

WITH BROWN BUTTER AND SAGE

Gnudi — or "naked" (as in ravioli) — are a variation on gnocchi, only these dumplings are made with ricotta rather than potatoes. This version is adapted from the iconic version that April Bloomfield serves at her New York City restaurant, The Spotted Pig.

**SERVES 4 AS AN APPETIZER OR
2 AS A MAIN COURSE**

1 pound fresh ricotta

3 ounces Parmesan, grated (about ¾ cup)

Kosher salt and freshly ground pepper

1 pound semolina flour (about 4 cups)

5 tablespoons unsalted butter

12 fresh sage leaves, cut into thin strips

1. Place the ricotta on a square of doubled cheesecloth. Gather the cheesecloth into a pouch and hang the pouch over a bowl to drain for 1 hour, kneading the bag occasionally to release additional whey. (To hang the pouch, tie it to a cabinet handle or suspend it from the faucet if your sink is deep enough.) Measure out 12 ounces of the drained ricotta. Reserve any extra ricotta for another use.

2. Combine the drained ricotta with ½ cup of the Parmesan in a medium bowl and season with salt and pepper. Chill in the freezer for 30 minutes (no longer; do not let mixture freeze).

Recipe continues on page 331

GNUDI

3. Spread the semolina in a 9-by-13-inch baking pan. Using a tablespoon, scoop a ball of ricotta mixture and drop it into the semolina. Gently rotate the pan to roll the ricotta ball and coat completely in semolina. Transfer the ricotta ball to a rimmed baking sheet. Repeat until all the ricotta mixture has been used.

4. Pour any remaining semolina over the gnudi and gently rotate the sheet to ensure the gnudi don't stick to the sheet or each other. Cover the sheet with plastic wrap and refrigerate for at least several hours or overnight.

5. To serve, bring a large pot of water to a boil. Melt the butter in a 12-inch skillet over medium-low heat. Let the butter cook until it starts to crackle and turns pale golden brown. Remove the skillet from heat.

6. Add gnudi and a generous pinch of salt to boiling water and cook for 3 minutes — they will float to the top. Remove the gnudi from the water with a slotted spoon and transfer to the browned butter in skillet. Add the sage and cook over medium heat, swirling pan occasionally, until gnudi are cloaked with browned butter, 1 to 2 minutes.

7. Transfer to serving plates and spoon any remaining brown butter and sage over the top. Top with remaining Parmesan and serve immediately.

SAAG PANEER
(SPINACH WITH PANEER)

Saag Paneer is a staple of Indian restaurant menus. You can easily make this version at home, and if there is freshly made paneer in the house, the work is almost complete.

SERVES 4

1 teaspoon extra-virgin coconut oil

1 medium yellow onion, thinly sliced

2–3 garlic cloves, minced

1 teaspoon cumin seeds

1 teaspoon fenugreek seeds

8 ounces paneer, cut into ½-inch cubes (about 2 cups)

8 cups chopped fresh spinach leaves (stemmed; from 2–3 bunches)

1–2 tablespoons crème fraîche or heavy cream

Kosher salt and freshly ground pepper

Lemon wedges, for serving

1. Heat oil in a large skillet over medium-high heat and add the onion, garlic, and cumin and fenugreek seeds. Cook, stirring occasionally, until the onion becomes tender, about 5 minutes.

2. Add paneer and fry, turning the cubes gently, until heated through, about 3 minutes.

3. Add spinach to the pan in batches, tossing until wilted before adding more (this should take about 3 minutes total). Stir in crème fraîche, adding enough so the mixture is not too dry, and cook to heat through, 1 minute. Season with salt and pepper and serve with lemon wedges.

Whey-Braised Pork Shoulder

Here's a good method for using leftover whey: it happens to have a terrific tenderizing quality when used to cook meat. Consider serving this pork dish with fluffy mashed potatoes — also enriched with whey instead of milk or cream — or try soft polenta.

SERVES 6

Boneless pork shoulder (3–4 pounds)

Kosher salt and freshly ground pepper

¼ cup olive oil

3 medium yellow onions, chopped (about 2½ cups)

4 garlic cloves, crushed

4 sprigs thyme

2 sprigs rosemary

2 dried bay leaves

3½ cups whey

1. Preheat the oven to 325°F (170°C). Dry the pork well with paper towels and season with salt and pepper. Heat the oil over medium-high heat in a large Dutch oven or other heavy pot. Add pork and sear until golden brown on all sides, about 4 minutes per side. Remove the pork and set aside to keep warm.

2. Remove all but 2 tablespoons of oil from the pot and add the onions, garlic, thyme, rosemary, and bay leaves. Sauté, stirring occasionally, until onions are tender and mixture is fragrant, about 8 minutes. Return the pork to the pot and add the whey. Bring to a simmer and cover.

3. Braise in the oven for 2–2½ hours, or until pork is nearly fork-tender. Remove lid and roast for 30 minutes longer. Transfer the pork to a carving board and let rest before slicing.

4. Remove herbs from the pot and discard. Transfer the liquid remaining in the pot to a blender and purée, being careful not to fill jar more than half-way each time (or use an immersion blender). Season sauce with salt and pepper.

5. Serve pork in thick slices with pan sauce spooned over the top.

Kachimoru
(Spiced Yogurt Curry)

Like all curries, this one is redolent of warm spices. It's also the creamy kind of curry that's enriched with yogurt (unlike other curries, which rely on coconut milk). The yogurt helps tame the heat of the chile peppers; removing the fresh chile's seeds also helps. Enjoy kachimoru spooned over basmati rice.

SERVES 4

8 ounces yogurt

½ cup water

1 teaspoon kosher salt

2 tablespoons extra-virgin coconut oil

1 teaspoon brown mustard seeds

½ teaspoon fenugreek seeds

1 green cardamom pod

1–2 dried red chiles

3 shallots, peeled and cut into thin slivers

1 piece (½ inch) peeled fresh ginger, cut into fine slivers

1 serrano chile or jalapeño, cut into thin rings (seeded if less heat is desired)

½ teaspoon ground turmeric

1. Whisk together the yogurt, water, and salt in a medium bowl.

2. Heat the oil over medium-high heat in a large skillet. Add the mustard and fenugreek seeds, cardamom, and dried chile (use less for a milder dish). Cook, stirring constantly, until seeds start to pop, about 2 minutes. Add the shallots, ginger, and serrano chile. Cook until fragrant and the onions start to soften, stirring occasionally, 7–10 minutes. Do not let the curry boil or it will curdle.

3. Stir in the turmeric. Remove from heat and stir in yogurt mixture (the sauce will be on the thin side). Serve hot.

Gözleme
(Stuffed Turkish Flatbreads)

These flatbreads make a filling meal on their own or a marvelous addition to a mezze spread. They can be filled with any assortment of ingredients; feel free to swap out the feta for halloumi, mizithra, kasseri, or other Turkish or Greek cheeses. Gözleme are never better than when fresh out of the skillet, but are also delicious at room temperature (as a great make-ahead party food).

MAKES 4

2¼ cups plus 2 tablespoons all-purpose flour

1 cup water

2 ounces yogurt

¼ teaspoon active dry yeast (from 1 packet)

Kosher salt

3 tablespoons olive oil, plus more for pan

1 small leek or white onion, finely chopped (leek washed well and drained)

2 ounces feta, crumbled (about ½ cup)

1 teaspoon dried mint or oregano, preferably Greek

1 teaspoon red pepper flakes (optional)

1. Put the flour, water, yogurt, yeast, and ½ teaspoon salt in a large bowl. Mix with a spoon to form a very wet dough. Cover and let proof for 10–15 minutes.

2. Lightly coat a work surface with 1 tablespoon of the oil and turn the flour mixture out onto the surface. Knead until it comes together into an elastic, not-too-sticky dough. Return dough to bowl, cover, and let rise in a warm, draft-free place for 1 hour.

3. Heat 1 tablespoon of the oil over medium-high heat in a medium skillet. Add the leek and sauté until tender and just starting to turn golden, about 5 minutes. Season lightly with salt. Remove from heat and let cool.

4. Coat a work surface with the remaining 1 tablespoon oil. Turn the dough out onto the surface and cut into four equal pieces. One at a time, roll each piece out into a very thin round, 9–10 inches in diameter. Sprinkle the center of one round with one-quarter of the leek, feta, mint, and pepper flakes (if using). Fold the edges of the dough up and over the filling to form a 6-inch square.

Recipe continues on next page

Press lightly with the rolling pin to seal the dough and stretch the flatbread out to a 7-inch square. Repeat with the remaining dough and filling.

5. Heat a large skillet over medium heat. Oil lightly and cook the gözleme until well browned and cooked through, about 3 minutes per side. Repeat until all gözleme are cooked. Cut into triangles or strips and serve hot or at room temperature.

Pizzoccheri
(Buckwheat Pasta with Fontina and Cabbage)

This is, hands down, the quintessential après-ski meal, courtesy of the Lombardy region of northern Italy. The flat pasta strands, with the nutty taste and toothsome texture of buckwheat flour, are what define this dish; fontina is the traditional cheese, though other Alpine cheeses are more than welcome here.

SERVES 4

2 tablespoons unsalted butter

1 pound napa or Savoy cabbage, shredded (about 3 cups)

1 small yellow onion, thinly sliced (about ¾ cup)

1 large russet potato, peeled and diced (about 1½ cups)

½ pound Buckwheat Noodles (recipe follows)

6 ounces fontina, cubed (about 1½ cups)

Kosher salt and freshly ground pepper

1. Melt the butter over medium heat in a large skillet. Sauté the cabbage and onion until they soften and start to turn golden, about 10 minutes. Remove skillet from heat.

2. Cook the potato in a large pot of boiling water until fork-tender, about 8 minutes. Add the noodles to the pot and cook until just tender, about 2 minutes for homemade noodles. Ladle out and reserve 2 cups pasta water.

3. Drain the potato and pasta and add to the skillet with the cabbage. Add ½ cup of the pasta water to the pan. Place the skillet over medium heat and cook, gently stirring, until the pasta and cabbage are coated with a silky sauce. If the pasta seems too dry, add a little more pasta water. Add the cheese to the skillet, and toss gently to combine. Season generously with salt and pepper and serve immediately.

Buckwheat Noodles

MAKES 1 POUND

2 cups fine buckwheat flour, sifted

¾ cup all-purpose flour, plus more for dusting

½ cup water

Pinch of kosher salt

1. Mix the buckwheat flour, ½ cup of the all-purpose flour, the water, and salt together in a medium bowl (or on a wooden board). Knead to form a firm dough. Let the dough rest for 20 minutes.

2. Turn dough out onto a lightly floured work surface and roll into ⅛-inch-thick sheets, or roll out in a pasta machine.

3. Cut pasta into ½-by-2-inch strips (by hand or with machine). Toss with remaining ¼ cup all-purpose flour and reserve on a baking sheet until ready to cook. The noodles can be frozen for up to 3 months; first place the baking sheet with noodles in freezer until firm, about 30 minutes, then transfer noodles to a resealable plastic bag (or divide in half). Do not thaw before cooking.

OVEN-FRIED ARANCINI
WITH BACON AND CHEDDAR

Arancini ("little orange" in Italian), American-style: Take classic risotto and flavor it with bacon and cheddar before forming into balls and "frying" in the oven for easy preparing. The rice is also a simplified version of more labor-intensive risotto, though you won't discern the difference. Feel free to veer toward tradition and make the recipe using pancetta and Montasio or fontina, or prosciutto and taleggio. In every which way the little morsels will be meltingly good.

MAKES ABOUT 2 DOZEN

4 ounces bacon

1 tablespoon extra-virgin olive oil,
    plus more for baking sheet

1 garlic clove, minced

4 scallions, thinly sliced

¾ cup arborio rice

3 ounces cheddar, shredded (about ¾ cup)

2 ounces Parmesan, grated (about ½ cup)

1 egg

    Kosher salt and freshly ground pepper

1 cup panko breadcrumbs

1. Cook bacon in a hot skillet until crisp, turning as needed; drain on paper towels, then crumble.

2. Preheat the oven to 450°F (230°C). Heat the oil in a medium saucepan over medium heat. Add the garlic and half the scallions, and sauté until softened, stirring occasionally, about 5 minutes.

3. Add the rice to the pan and cook, stirring occasionally, until toasted, 1 or 2 minutes. Pour in 1½ cups water and bring to a boil. Immediately reduce the heat, cover the pan, and simmer, stirring occasionally, until the water is absorbed into the rice. Spread the rice onto a rimmed baking sheet and let cool to room temperature.

4. Stir the bacon, cheddar, half the Parmesan, the egg, and remaining scallions into the rice. Season with salt and pepper to taste.

5. Mix the panko with the remaining Parmesan in a bowl. Dampen hands with water and roll the rice into 1½- to 2-inch balls. Dredge each ball in panko mixture. Place rice balls onto a lightly greased baking sheet.

6. Bake arancini for 20–30 minutes, until golden brown. Serve hot.

SHRIMP SAGANAKI
(SHRIMP BAKED WITH FETA AND TOMATOES)

Named for the traditional small pan used to prepare this Greek dish, saganaki is most famously made with crisp fried cheese flamed with ouzo. Shrimp saganaki may not have the same fiery flamboyance, but it is still worthy of appreciation. Be sure to have bread nearby for soaking up every last bit of the luscious broth left behind.

SERVES 4

- 3 tablespoons extra-virgin olive oil
- 1 fennel bulb, trimmed (reserve a few fronds for garnish) and thinly sliced (about 1½ cups)
- 1 small yellow onion, halved and thinly sliced (about ¾ cup)
- 1 teaspoon anise seed
- 2 large tomatoes, cored and chopped (about 2 cups)
- Kosher salt and freshly ground pepper
- 1 pound medium shrimp, peeled and deveined
- 4 ounces feta or halloumi, crumbled (about 1 cup)
- 1 lemon, cut into wedges

1. Preheat the oven to 450°F (230°C). Heat 2 tablespoons of the oil in an 8-inch cast-iron or other oven-safe skillet over medium-high heat. Sauté the fennel, onion, and anise seed until the onion and fennel are tender, 5–8 minutes. Add the tomatoes and bring to a simmer. Season with salt and pepper.

2. Add the shrimp to the skillet and sprinkle the feta over the top.

3. Transfer the skillet to the oven and bake for 6–8 minutes, until shrimp are opaque throughout. Serve warm, garnished with reserved fennel fronds, lemon wedges, and remaining 1 tablespoon oil.

SHRIMP SAGANAKI

16

DESSERTS

SANDESH

Sandesh

These no-bake Bengali sweets are an excellent reason to make a batch of paneer — or to use up what you already have on hand. Ground cardamom gives the melt-in-your-mouth treats their characteristic spice; almonds, pistachios, and saffron are added here for even more authentic flavors. While optional, the milk powder helps to firm up the sandesh, giving them a chewier texture and making them easier to unmold.

MAKES 1 DOZEN

1 pound paneer

1 tablespoon dry milk powder (optional)

½ cup confectioners' sugar, plus more for dusting

½ teaspoon ground cardamom

4 tablespoons melted unsalted butter or ghee

12 saffron threads

2 tablespoons slivered almonds

2 tablespoons chopped pistachios

1. Using a potato masher or a wooden spoon, mix the paneer and milk powder, if desired, in a medium saucepan. Cook over medium heat until the paneer softens, stirring frequently, about 2 minutes. Remove from heat and cool slightly.

2. Turn the paneer out onto a lightly sugared work surface. Knead in the sugar and cardamom until a smooth paste is formed.

3. Grease 12 cups of a muffin tin with the melted butter. Divide the saffron threads among the cups. Sprinkle an equal amount of almonds and pistachios into each cup. Divide the dough among the cups, pressing firmly into the molds and flattening the tops.

4. Chill for at least 4 hours, preferably overnight. Use a butter knife or thin-bladed spatula to ease the sandesh out of the cups to serve.

Plum and Ricotta Galette

Despite its fancy French name, a galette is actually among the simplest of tarts to make — no fitting into a pie plate, crimping, and trimming required. This free-form dessert features a rye crust, for unexpected flavor, and a filling that's just right for bridging summer to fall, when plums are plentiful and at their flavorful best. Ricotta adds welcome creaminess, both in the baked filling layer and as a topping when serving.

MAKES 1 TART

1 cup rye flour

⅔ cup all-purpose flour

3 tablespoons granulated or coconut sugar

½ teaspoon salt

½ cup (1 stick) plus 2 tablespoons cold unsalted butter, cut into ½-inch cubes

¼–½ cup ice water

1 cup ricotta, plus more for serving

2 eggs, separated

Zest of 1 lemon, plus more for serving

Pinch of ground cinnamon

6 plums, pitted and cut into thin wedges

¼ cup turbinado or demerara sugar, for sprinkling

Recipe continues on next page

1. Combine ⅔ cup of the rye flour, the all-purpose flour, 1 tablespoon of the granulated sugar, and the salt in a food processor. Pulse 2 or 3 times. Scatter ½ cup of the butter across the flour mixture. Pulse until mixture forms coarse crumbs. Add ¼ cup ice water and pulse to combine into a dough. Add more ice water, 1 tablespoon at a time, if dough is too dry. Remove from processor, wrap in plastic wrap, and chill for at least 1 hour or up to overnight.

2. Whisk the ricotta, egg yolks, remaining 2 tablespoons granulated sugar, the lemon zest, and cinnamon in a medium bowl.

3. Line a baking sheet with parchment paper. Use the remaining ⅓ cup rye flour to roll out the chilled crust into a 12-inch round. Lay crust onto the lined baking sheet. Leaving a 1½-inch border, spread ricotta mixture evenly over the crust. Evenly arrange the plums over the ricotta filling. Scatter the remaining 2 tablespoons butter over the plums. Gently lift the crust up and over the plums, pleating the crust neatly.

4. Make an egg wash by beating the egg whites with 1 tablespoon of water. Brush over exposed crust, then sprinkle entire tart liberally with turbinado sugar. Chill tart for 30 minutes.

5. Preheat the oven to 400°F (200°C), with rack on lowest shelf. Bake tart for 30 to 40 minutes, or until crust is browned and crisp and the plums are juicy. Let cool slightly before serving, dolloped with more ricotta and sprinkled with lemon zest, if desired.

FROMAGE BLANC CAKE

As if fromage blanc wasn't delicious enough on its own, the French found a way to impart its richness to a loaf cake that's the equivalent of old-fashioned pound cake. Meaning it's excellent on its own or topped with peaches and cream — make that crème fraîche — or myriad other combinations of fruit and creamy goodness.

MAKES 1 LOAF

2 tablespoons unsalted butter, softened, plus more for pan

1½ cups all-purpose flour

2 teaspoons baking powder

¼ teaspoon salt

1 cup sugar

Zest of 2 lemons

½ cup fromage blanc

3 eggs

½ cup extra-virgin olive oil

1. Preheat the oven to 350°F (180°C). Grease a 9-by-5-inch loaf pan with butter. Whisk the flour, baking powder, and salt in a medium bowl.

2. Put the sugar and zest in a medium bowl. Whisk in the fromage blanc and then the eggs, one at a time, until mixture is smooth.

3. Whisk the flour mixture into the fromage blanc mixture. Stir in the oil. Scrape the batter into the prepared pan and smooth the top.

4. Bake for 40–50 minutes, or until the cake is golden brown and starts to pull away from the sides of the pan. Let cool for 10 minutes in pan

on a wire rack, then turn cake out onto rack and let cool completely before serving.

NOTE: This cake keeps exceptionally well in the freezer, wrapped well in plastic wrap, for up to 3 months; thaw at room temperature to serve.

APPLE PIE
WITH AGED-CHEDDAR CRUST

Melting a slab of cheddar over a slice of apple pie is a common way to order this all-American favorite combination in diners across the land. Here, the cheese is grated and mixed into the dough; for the best results, use an aged cheddar. And by all means: Top each slice with more cheddar and run under the broiler, watching carefully until just melted.

MAKES ONE 9-INCH PIE

- 2½ cups plus 2 tablespoons all-purpose flour, plus more for dusting
- ¾ teaspoon salt
- 1 cup (2 sticks) plus 2 tablespoons cold unsalted butter, cut into ½-inch cubes
- 8 ounces aged cheddar, grated (about 2 cups)
- ¼–½ cup ice water
- 2½ pounds mixed apples, such as Cortland, Braeburn, and Granny Smith, peeled, cored, and cut into wedges (about 8 cups)
- ¾ cup sugar

1. Combine 2½ cups of the flour, ¼ teaspoon of the salt, and 1 cup of the butter in a food processor. Pulse until mixture resembles coarse crumbs. Add 1 cup of the cheese to the processor bowl and ¼ cup of the ice water and pulse just until dough comes together. If a dough does not form, add additional water, 1 tablespoon at a time.

2. Divide dough in half. Shape each piece into a disk and wrap in plastic wrap. Chill for at least 1 hour or overnight.

3. Preheat the oven to 425°F (220°C). Combine the apples, sugar, and the remaining 1 cup cheese, 2 tablespoons flour, and ½ teaspoon salt in a large bowl. Mix together to coat the apples evenly.

4. On a lightly floured surface, roll one disk of dough out into a 12-inch round. Fit dough into a 9-inch pie plate. Fill with the apple mixture, mounding it into the center of the dough.

5. Roll the second disk of dough out into a 12-inch round. Place over pie plate and crimp the edges of top and bottom doughs to seal. Cut six slits into the top dough.

6. Place pie plate on a rimmed baking sheet to catch any dripping juices. Bake for 25 minutes.

7. Reduce oven heat to 350°F (180°C) and bake for 40 additional minutes, or until apples are tender and juices are bubbling; cover loosely with foil if crust is browning too quickly. Let cool before serving; pie is best when allowed to rest overnight (otherwise it can be runny). Reheat in a low oven if desired before serving.

NOTE: The dough can be frozen, in ziplock freezer bags, for up to 3 months; thaw in the refrigerator before using.

Knafeh
(Middle Eastern Feta Pie)

Knafeh is a Lebanese dessert made with nabulsi, a brined cheese that's squeaky and firm at room temperature, yet gets oozy and stretchy when melted. It's often described as a cross between fresh mozzarella and feta, both of which are used here to try and replicate nabulsi's singular texture. This new take on knafeh also includes dried fruit, for welcome sweetness.

Kataifi pastry is akin to phyllo dough and needs to be well buttered to turn wonderfully golden brown and crisp in the oven. Look for kataifi in the freezer section of specialty food stores and Middle Eastern markets, or buy shredded phyllo instead.

SERVES 12

- 1 cup (2 sticks) unsalted butter, melted, plus more for baking dish
- 1 pound frozen kataifi or shredded phyllo, thawed according to package instructions
- 8 ounces fresh mozzarella, shredded (about 2 cups)
- 8 ounces feta, crumbled (about 2 cups)
- 1 cup golden raisins
- 1 cup heavy cream, whipped
- 1 cup sugar
- 1 (3-inch) strip orange peel
- 1 (3-inch) strip lemon peel
- 1 cinnamon stick
- Pinch of salt

1. Preheat oven to 400°F (200°C). Grease a 9-by-13-inch glass baking dish. Combine the kataifi and butter in a large bowl. Gently lift and pull the kataifi strands apart to coat each one well. Layer half the kataifi evenly in bottom of prepared dish.

2. Spread the cheeses and raisins evenly over the kataifi, then dot evenly with the whipped cream. Evenly layer remaining kataifi over the top, compressing it gently.

3. Bake for 30 minutes, or until the pastry is a dark golden brown; cover loosely with foil if it browns too quickly.

4. Meanwhile, combine the sugar, ½ cup water, citrus peels, cinnamon, and salt in a small saucepan. Bring to a boil, stirring until sugar is dissolved, then reduce heat and simmer until syrup is starting to thicken, about 10 minutes. Carefully remove the citrus and cinnamon. Pour the hot syrup over the hot pastry. Let the knafeh cool for about 10 minutes before cutting into pieces and serving, warm or at room temperature.

NOTE: Leftovers can be stored in an airtight container for a few days; reheat in a 250°F (120°C) oven if desired.

Tea Tiramisu

This modern take on the age-old Italian favorite replaces the usual espresso with Earl Grey tea and incorporates plenty of citrus for a brighter result. Look for ladyfingers that are light for their size — they'll absorb lots of flavorful tea syrup. Because the dessert rests on the quality of the mascarpone, it's worth making your own; see page 64 for the recipe.

SERVES 8

- 2 medium oranges
- 2 medium lemons
- 6 Earl Grey tea bags
- 2 cups boiling water
- ¾ cup sugar
- 4 egg yolks
- ⅓ cup brandy
- ¼ teaspoon salt
- 1 pound mascarpone
- 1 cup heavy cream, chilled
- 36 ladyfingers, preferably Italian

1. Zest the oranges and lemons and reserve the zest. Juice the lemons and measure out ¼ cup of juice; reserve the rest for another use. Peel the oranges and cut the fruit into thin rounds or segments. Reserve for garnishing the tiramisu.

2. Put the tea bags in a small bowl and pour the boiling water over them. Let steep for 5 minutes. Remove tea bags. Add the lemon juice and ¼ cup of the sugar. Stir until the sugar melts.

3. Combine the orange and lemon zests, remaining ½ cup sugar, the egg yolks, brandy, and salt in a heatproof bowl set over a pan of simmering water. Whisk (by hand or electric mixer) until tripled in volume. Remove from heat. Whisk in mascarpone.

4. Beat the cream in a bowl until it can hold a stiff peak. Using a flexible spatula, fold in mascarpone mixture until evenly incorporated.

5. Dip half of the ladyfingers into the sweet tea mixture. Line the bottom of a 9-by-13-inch baking pan with the soaked ladyfingers. Spread one-half of the mascarpone cream over the top. Repeat with the remaining ladyfingers and mascarpone cream.

6. Top the tiramisu with orange slices. Chill at least 6 hours or overnight before serving.

KÄSEKUCHEN

Käsekuchen

Every culture has its cheesecake, from crustless Italian ricotta cheesecake to cookie-crust New York–style cheesecake. Käsekuchen (Käse means cheese and Kuchen means cake or dessert) is the German iteration of the universally adored dessert. Traditional versions are made with quark, and that's what you will find here. Don't be tempted to substitute yogurt or cream cheese, as the result just won't be the same. Garnish with fresh berries and dust with confectioners' sugar, if desired.

MAKES ONE 10-INCH CAKE

- 2 cups plus 3 tablespoons all-purpose flour
- ½ cup plus 2 tablespoons sugar
- ¼ teaspoon salt
- ½ cup (1 stick) chilled unsalted butter, cut into cubes
- 6 tablespoons unsalted butter, melted
- 6 eggs, separated
- 2 tablespoons water
- 1½ pounds quark
- 1 teaspoon vanilla extract
- Zest and juice of 1 medium lemon (about 3 tablespoons juice)

1. Purée 2 cups of the flour, 2 tablespoons of the sugar, and ⅛ teaspoon of the salt in a food processor to combine. Add the cold butter cubes and pulse until mixture forms coarse crumbs. Add 2 of the egg yolks and the water and pulse to combine into a dough. Remove from processor, wrap in plastic wrap, and chill for at least 1 hour.

2. Preheat the oven to 350°F (180°C). Grease a 10-inch springform pan with 2 tablespoons of the melted butter.

3. Use the remaining 3 tablespoons flour to roll the crust out into a 15-inch circle. Transfer crust to pan and press it into the bottom and up the sides of the pan. Chill for 1 hour.

4. Combine the quark, remaining ½ cup sugar, remaining 4 egg yolks, vanilla, lemon zest and juice, and the remaining ⅛ teaspoon salt in a food processor. Pulse until well blended and smooth. Pulse in the remaining 4 tablespoons melted butter.

5. Beat 4 of the egg whites in a large bowl with an electric mixer at medium-high speed until stiff peaks form. (Reserve the remaining 2 egg whites for another use.) Gently fold the beaten egg whites into the quark mixture. Carefully pour the mixture into the springform pan.

6. Bake for 45–60 minutes, or until the top is lightly browned and the filling is set (test by jiggling; the middle should no longer wobble).

7. Open the oven door and let cake cool in the oven for 10 minutes. Run a knife around the side of the pan to loosen the crust. Release and remove springform sides and let the cake cool completely before serving.

Goat Cheese and Honey Ice Cream

It's in, it's out, it's in, it's out: I say goat cheese ice cream is most decidedly in — and deserves a permanent spot in the dessert diaspora. Honestly, the taste of ice cream made with goat cheese is a one-of-a-kind experience. The ricotta and crème-fraîche variations are also beyond compare.

MAKES 1 QUART

1½ cups heavy cream

4 ounces goat cheese, at room temperature

4 egg yolks

1½ cups whole or low-fat milk

½ cup honey

⅛ teaspoon salt

1. Whisk the heavy cream and goat cheese together in a medium bowl until smooth.

2. Put the egg yolks in another medium bowl and whisk (by hand or electric mixer) until lightened, about 2 minutes.

3. Combine the milk, honey, and salt in a medium saucepan and heat until just steaming. Do not let the mixture boil or the milk might curdle. Whisking constantly, pour the warm milk mixture slowly into the yolks to temper. Pour this mixture back into the saucepan and heat over medium-low heat, stirring with a heatproof flexible spatula, until an instant-read thermometer registers 165°F (74°C) and the mixture is thick enough to coat the back of a spoon. This will take about 10 minutes.

4. Strain goat cheese mixture through a fine sieve into a medium bowl. Add the egg mixture and whisk well to combine. Chill overnight.

5. Freeze in an ice-cream maker according to manufacturer's instructions. Transfer to an airtight container and freeze up to 2 weeks; let soften at room temperature for a few minutes before scooping and serving.

Variations

Ricotta Rum–Raisin

Substitute ricotta cheese for the goat cheese and add ½ teaspoon ground cinnamon to the milk mixture in step 3; fold ½ cup rum-soaked raisins into the ice cream before churning.

Crème Fraîche and Brandied Cherry

Substitute crème fraîche for the goat cheese and ¾ cup packed brown sugar for the honey; fold ½ cup chopped brandied cherries and ¼ cup chopped toasted almonds into the ice cream before churning.

BUTTERMILK COOKIES

WITH LEMON GLAZE

Buttermilk makes these cookies delightfully tender and slightly tangy, with the bright taste of lemon zest in the cookies and the buttermilk glaze. You could just as easily make them with lime or orange zest, depending on what you have on hand. They're ideal nibbled with a cup of hot tea or espresso.

MAKES ABOUT 5 DOZEN

FOR THE COOKIES

3 cups all-purpose flour

1 teaspoon finely grated lemon zest

½ teaspoon baking soda

½ teaspoon salt

¾ cup (1½ sticks) unsalted butter, softened

1¼ cups granulated sugar

2 eggs

1 teaspoon vanilla extract

⅔ cup buttermilk

FOR THE GLAZE

1½ cups confectioners' sugar

3 tablespoons buttermilk

½ teaspoon vanilla extract

2 teaspoons finely grated lemon zest

1. Preheat the oven to 350°F (180°C) and line two baking sheets with parchment paper. Whisk the flour, zest, baking soda, and salt together in a medium bowl.

2. Beat the butter and granulated sugar in a large bowl with an electric mixer on medium-high speed until pale and fluffy, about 5 minutes. Add eggs, one at a time, beating well after each addition and scraping down sides of bowl as needed. Beat in vanilla.

3. With mixer on low speed, add the flour mixture in three batches, alternating with the buttermilk and beginning and ending with flour; beat just until smooth and combined.

4. Drop level tablespoons of dough onto the prepared baking sheets, about 2 inches apart. Bake one sheet at a time for 12–15 minutes, until cookies are slightly puffed in center and just starting to turn golden around the edges. Let cookies cool on the sheet for 1 minute before transferring to wire racks.

5. While cookies are baking, make the glaze: Combine the confectioners' sugar, buttermilk, vanilla, and zest in a medium bowl and whisk until smooth.

6. Brush glaze onto tops of warm cookies. Let cookies cool completely before serving. Store between layers of parchment in airtight containers at room temperature for up to 1 week.

GLOSSARY

Acidity. The amount of sourness in milk. Acidity is produced by the activity of starter culture bacteria, and it precipitates the milk protein into a solid curd.

Aging. A step in cheese making in which the cheese is stored at a particular temperature and relative humidity for a specified amount of time so it can develop its distinct flavor.

Albuminous protein. Protein in milk that cannot be precipitated out by the addition of rennet. Albuminous protein remains in the whey and is precipitated by high temperatures to make ricotta.

Annatto. A natural vegetable extract added to milk prior to renneting to impart various shades of yellow to cheese. Obtained from the seeds of the fruit of the annatto tree, *Bixa orellana*, which is native to many tropical countries and cultivated in the West Indies, Brazil, and India.

Ash. Charcoal derived from vegetable matter, used to neutralize the surface pH of cheese and create a friendly environment for mold growth.

Bacteria. Microscopic unicellular organisms found almost everywhere. Lactic acid–producing bacteria are helpful and necessary for making quality hard cheeses.

Bacteria-ripened cheese. A cheese upon whose surface bacterial growth is encouraged to develop in order to produce a distinct flavor. Brick and Limburger are examples of bacteria-ripened cheeses.

Bandaging. Rubbing cheese with lard and wrapping it with butter muslin to keep its shape, preserve its coat, and prevent the loss of excess moisture through evaporation.

Blue mold. See *Penicillium roqueforti*.

Brevibacterium linens. A red bacterium that is encouraged to grow on the surfaces of certain cheeses, such as brick and Limburger, to produce a sharp flavor.

Butterfat. The fat portion (cream) of milk, expressed as a percentage of total weight. Butterfat content varies depending on type of animal the milk comes from, the diet of the animal, and other factors.

Butter muslin. A very tightly woven cotton cloth used to drain soft cheeses.

Calcium chloride. A white, lumpy solid that is derived from calcium carbonate and used as a drying agent to produce a firmer curd.

Casein. The principal protein in milk that, with the addition of rennet, coagulates and forms the foundation for cheese.

Cheese board. A small piece of ash, fir, maple, or birch with no knots, typically measuring 6 inches square and 1 inch thick, often used to help drain soft cheeses, such as Camembert. Larger cheese boards are often used to hold cheeses during the aging process.

Cheesecloth. A finely woven cotton cloth used to drain curds, line cheese molds, and perform a host of other cheese-making functions. Use only cheesecloth intended for use in cheese making.

Cheese color. See annatto.

Cheese mat. A piece of wooden reed or food-grade plastic, often used to help drain soft cheeses, such as Coulommiers and Camembert.

Cheese mold. A form made of food-grade plastic or stainless steel, into which curds are placed for draining and/or pressing. The cheese mold helps produce the final shape of the cheese and aids in drainage.

Cheese salt. A coarse, noniodized flake salt.

Cheese starter culture. A quantity of live bacteria added to milk as the first step in making many cheeses. The bacteria produce an acid during their life cycle in the milk. There are two categories of starter culture: mesophilic and thermophilic.

Cheese trier. A stainless-steel tool used to take samples from a cheese as it ages to determine whether it is properly matured and ready for eating.

Cheese wax. A pliable, low-melting-point wax that creates an airtight seal that will not crack. Most hard cheeses are waxed.

Cheese wrap. A permeable cellophane for wrapping cheese.

Clean break. Response of the curd to testing that shows it is ready for cutting. The curd has reached the desired stage when the tip of a knife inserted slightly at a 45-degree angle separates the curd firmly and cleanly.

Coagulation. The point at which milk congeals into a thickened mass.

Cooking. A step in cheese making during which the cut curd is warmed to expel more whey.

Coulommiers mold. A stainless-steel mold consisting of two hoops, one resting inside the other. The mold is used to make Coulommiers cheese. (A one-piece Camembert mold may also be used.)

Curd. The solid, custardlike state of milk achieved by the addition of rennet. The curd contains most of the milk protein and fat.

Curd knife. A flat, thin knife with a blade long enough to reach the bottom of the pot without immersing the handle and with a rounded end rather than a pointed one.

Cutting the curd. A step in cheese making in which the curd is sliced into equal-size pieces.

Dairy thermometer. A thermometer with a range from 0°F to 220°F (−18 to 104°C), used to measure the temperature of milk during cheese making.

Direct heating. Heating milk for cheese making on the stovetop.

Direct-set starter. A prepackaged starter culture added directly to milk to turn milk protein into a solid white gel for the purpose of cheese making.

Draining. A step in cheese making in which the whey is separated from the curds by ladling the curds and whey into a colander lined with cheesecloth.

Drip tray. A receptacle placed under a mold while a cheese is pressed. The drip tray has a spout to drain the whey into a sink or other container.

Drying. Phase of cheese making during which a cheese is allowed to sit and evaporate moisture, so as to form a protective rind in anticipation of the aging process.

Dry milk powder. Dehydrated milk solids that may be reconstituted in water.

Flora Danica starter. A fresh goat-cheese starter that may be used as either a direct-set or a reculturable starter.

Follower. A disk of stainless-steel, food-grade plastic, or wood, placed between the curd in a mold and a weight; used to apply even pressure to the curd.

Geotrichum candidum. A white mold that is encouraged to grow on the surface of a number of soft mold-ripened cheeses, such as Camembert and Brie, producing a delicious, mottled white skin.

Hard cheese. A cheese that is firm or hard in texture because a high percentage of moisture is removed during the cheese-making process. Hard cheeses are pressed and aged for varying lengths of time for full flavor development.

Homogenization. A mechanical process that breaks up fat globules in milk so that they will be evenly dispersed and the cream will no longer rise to the top.

Indirect heating. Heating milk for cheese making by placing the pot into a large container or sink full of hot water.

Junket rennet. A weak form of rennet.

Lactic acid. The acid created in milk during cheese making. Cheese starter culture bacteria consume the milk sugar (lactose) and produce lactic acid as a by-product.

Lactose. The sugar naturally present in milk. Lactose can constitute up to 5 percent of the total weight of milk.

Lipase powder. An enzyme added to milk to create a stronger-flavored cheese.

Low-fat milk. Milk that has most of the cream removed, resulting in a butterfat content of 1 to 2 percent.

Mesophilic cheese starter culture. A blend of lactic acid–producing bacteria that is used to make cheeses when the cooking temperature is 102°F (39°C) or lower.

Milkstone. A milk residue that is deposited on cheese-making utensils over time.

Milling. A step in cheese making during which the curd is broken into smaller pieces before being placed in a cheese press.

Molding. A step in cheese making during which the curd is placed into a cheese mold.

Mold-ripened cheese. A cheese upon whose surface and/or interior a mold is encouraged to grow. The two types of mold most commonly used are blue mold (for blue cheeses) and white mold (for Camembert and related cheeses). See also *Penicillium roqueforti* and *Penicillium camemberti*.

Noncorrosive. A material that will not break down or erode in the presence of acid.

Oiling. The application of vegetable oil to provide a protective layer to keep a cheese from drying out.

Pasteurization. The heating of milk to 145°F (63°C) for 30 minutes in order to destroy pathogenic organisms that may be harmful to people.

Penicillium camemberti. A white mold that is encouraged to grow on the surface of a number of soft mold-ripened cheeses, such as Camembert and Brie, producing a delicious, mottled white skin.

Penicillium candidum. A white mold that is encouraged to grow on the surface of a number of soft mold-ripened cheeses, such as Brie and Camembert, producing a delicious, mottled white skin.

Penicillium roqueforti. A blue mold that is encouraged to grow on the surface and in the interior of a variety of blue cheeses, including Stilton and Gorgonzola.

Prepared starter culture. A culture of live bacteria used to acidify milk during cheese making. Kept properly, it can be propagated to create many generations of bacteria.

Pressing. A step in cheese making during which the cooked curds are placed into a cheesecloth-lined mold with weight on top of them to force out more whey.

Proper break. Desired result of a test made during the making of Swiss cheese. To make certain the curds are properly cooked, a handful is wadded into a ball. If the ball of curds can be easily broken back into individual curd particles, that is considered a proper break.

Raw milk. Milk that has been freshly taken from an animal and has not been pasteurized.

Reculturable starter. See prepared starter culture.

Red bacteria. See *Brevibacterium linens*.

Re-dressing. The changing of cheesecloth on a cheese that is being drained or pressed. This is required to keep the cheesecloth from sticking to the cheese.

Rennet. A substance that contains the enzyme rennin, which has the ability to coagulate milk. Rennet comes in liquid, tablet, paste, and powder forms. Animal-derived rennet comes from the fourth stomach of a milk-fed calf. Vegetable-derived rennet is a microbial rennet that contains no animal products.

Renneting. A step in the cheesemaking process in which rennet is added to milk to bring about coagulation.

Ripening. A step in cheese making in which milk is allowed to undergo an increase in acidity as a result of the activity of cheese starter culture bacteria.

Salting. The addition of cheese salt to the curds before molding, or the application of cheese salt to the surface of the finished cheese.

Saturated brine solution. A saltwater solution in which cheese is soaked. Water is saturated when it will no longer dissolve any additional salt.

Soft cheese. A cheese that is not pressed, has a high moisture content, and is either not aged at all or aged for a comparatively short period.

Thermophilic cheese starter culture. A bacterial starter culture that is used to make cheeses that have a high cooking temperature. Recipes for Italian and Swiss cheeses call for thermophilic culture.

Top-stirring. Stirring the top ¼ inch of nonhomogenized milk immediately after rennet has been added in order to keep the cream from rising.

Ultrapasteurization (UP). The high-heat treatment of milk to guarantee a long shelf life.

Waxing. Coating the outside of a cheese with cheese wax to keep it from drying out and to retard the growth of mold.

Whey. The liquid portion of milk that develops after the milk protein has coagulated. Whey contains water, milk sugar, albuminous proteins, and minerals.

White mold. See *Penicillium candidum, P. camemberti,* and *Geotrichum candidum*.

WHAT WENT WRONG?

PROBLEM
The cheese tastes very bitter.

Possible Causes and Solutions

Poor hygiene was used in handling the milk and/or cheese-making utensils. Make sure milk is kept cold and in sanitary conditions and properly sterilize all utensils (see page 30).

Too much rennet was used.

Excessive acidity developed during the cheese-making process. Learn about acid production on pages 36–38 and 55.

Too little salt was added to the curd after milling.

PROBLEM
The cheese tastes quite sour and acidic.

Possible Causes and Solutions

The cheese contains too much moisture. To reduce the moisture content during cheese making, see pages 42–44.

Excessive acidity developed during the cheese-making process. Learn about acid production on pages 36–38 and 55.

PROBLEM
The cheese has little or no flavor.

Possible Causes and Solutions

The cheese was not aged long enough. Follow directions in recipe for at least the minimum aging period.

Insufficient acidity was produced during cheese making. Learn about acid production on pages 36–38 and 55.

PROBLEM
After adding rennet, the milk almost instantly coagulates into a curd of tiny grains while the rennet is still being stirred into the milk.

Possible Causes and Solutions

There is excessive acidity in the milk. The milk should not start to coagulate until about 5 minutes after adding the rennet. Learn about acid production on pages 36–38 and 55.

PROBLEM
The milk does not coagulate into a solid curd.

Possible Causes and Solutions

This can be the result of using too little rennet or a poor-quality one. Make sure rennet is stored properly.

Rennet activity can be destroyed if it is diluted in overly warm water or contaminated with cheese coloring.

Rennet needs a particular temperature to set properly. Check the accuracy of your dairy thermometer by holding it in boiling water and making sure it reads 212°F (100°C).

Do not use milk that contains colostrum.

PROBLEM
The finished cheese is excessively dry.

Possible Causes and Solutions

An insufficient amount of rennet was added.

The curd was cut into particles that were too small.

The curds were cooked to an excessive temperature.

The curds were overly agitated.

PROBLEM
Mold growth occurs on the surface of air-drying or waxed cheese.

Possible Causes and Solutions

There were unclean aging conditions and/or the humidity was too high in the aging room.

PROBLEM
The cheese is quite difficult to remove from the mold after pressing.

Possible Causes and Solutions

Coliform bacteria and/or wild yeast may have contaminated the milk and the curd, producing gas that swells the cheese during pressing. Pay strict attention to cleanliness and proper storage of milk.

PROBLEM
The cheese, when cut open, is filled with tiny holes, giving it the appearance of a sponge.

Possible Causes and Solutions

Coliform bacteria and/or wild yeast contaminated it. Such contamination should be noted during the cooking process, as the curds will have an unusual odor similar to that of bread dough. Discard any contaminated curds or cheese.

PROBLEM
The cheesecloth is difficult to remove from the cheese after pressing. Pieces of the cheese rip off when the cheesecloth is removed.

Possible Causes and Solutions

Coliform bacteria and/or wild yeast have contaminated the cheese (see above).

The cheese was not re-dressed in a fresh cheesecloth when needed. This is particularly true for cheeses made with thermophilic starter cultures. Follow directions for each recipe.

PROBLEM
The cheese becomes oily when air-drying.

Possible Causes and Solutions

The cheese is being air-dried at too high a temperature. The temperature should not exceed 65°F.

The curd was stirred too vigorously or was heated to too high a temperature.

PROBLEM
Spots of moisture on the surface of the aging cheese beneath the wax begin to rot and ruin the cheese.

Possible Causes and Solutions

The cheese was not turned often enough. Turn the cheese at least daily when it first starts to age.

The cheese contains excessive moisture. To reduce the moisture content during cheese making, see pages 42–44, 51, and 56.

PROBLEM
There is insufficient acidity during cheese making.

Possible Causes and Solutions

The starter culture is not working, which could have several causes:

- Antibiotics are present in the milk. Check the source.

- The starter is contaminated. Use a new starter.

- There is cleaning agent residue, particularly chlorine, on the utensils. Rinse all utensils thoroughly.

PROBLEM
There is excessive acidity during cheese making.

Possible Causes and Solutions

The milk was improperly stored prior to cheese making or pasteurization. Store milk at 35°F (2°C) or lower until ready for cheese making.

Too much starter was added.

The ripening period was too long.

There is excessive moisture in the cheese. To reduce the moisture content during cheese making, see pages 42–44, 51, and 56.

PROBLEM
There is excessive moisture in the cheese.

Possible Causes

Acid development was inadequate during cheese making. Learn about acid production on pages 36–38 and 55.

The milk had too high a fat content. Butterfat content of milk should not be much higher than 4.5 percent.

The curds were heated too rapidly. Too fast an increase in cooking temperature produces a membrane around the curd particles that prevents moisture from escaping. Do not heat the curd faster than 2 degrees every 5 minutes.

Too much whey was retained in the curd. Cut the curd into smaller pieces.

The curd was heated to too low a temperature during cooking. Heat the curd to a somewhat higher temperature.

METRIC CONVERSION CHARTS

Unless you have finely calibrated measuring equipment, conversions between US and metric measurements will be somewhat inexact. It's important to convert the measurements for all of the ingredients in a recipe to maintain the same proportions as the original.

WEIGHT

To convert	to	multiply
ounces	grams	ounces by 28.35
pounds	grams	pounds by 453.5
pounds	kilograms	pounds by 0.45

US	Metric
0.035 ounce	1 gram
¼ ounce	7 grams
½ ounce	14 grams
1 ounce	28 grams
1¼ ounces	35 grams
1½ ounces	40 grams
1¾ ounces	50 grams
2½ ounces	70 grams
3½ ounces	100 grams
4 ounces	112 grams
5 ounces	140 grams
8 ounces	228 grams
8¾ ounces	250 grams
10 ounces	280 grams
15 ounces	425 grams
16 ounces (1 pound)	454 grams

VOLUME

To convert	to	multiply
teaspoons	milliliters	teaspoons by 4.93
tablespoons	milliliters	tablespoons by 14.79
fluid ounces	milliliters	fluid ounces by 29.57
cups	milliliters	cups by 236.59
cups	liters	cups by 0.24
pints	milliliters	pints by 473.18
pints	liters	pints by 0.473
quarts	milliliters	quarts by 946.36
quarts	liters	quarts by 0.946
gallons	liters	gallons by 3.785

US	Metric
1 teaspoon	5 milliliters
1 tablespoon	15 milliliters
¼ cup	60 milliliters
½ cup	120 milliliters
1 cup	230 milliliters
1¼ cups	300 milliliters
1½ cups	360 milliliters
2 cups	460 milliliters
2½ cups	600 milliliters
3 cups	700 milliliters
4 cups (1 quart)	0.95 liter
4 quarts (1 gallon)	3.8 liters

HOMEMADE CHEESE RECORD FORM

Keep track of your home cheese-making efforts. Visit https://www.storey.com/cheese-record-form/ to download a printable form for documenting your cheese-making process, tracking your cheese's progress, noting what you'd try differently, and recording your successes.

HOMEMADE CHEESE RECORD FORM

TYPE OF CHEESE

Date Made . . . .   Type of Milk . . . .   Amount of Milk . . . .

1. RIPENING

Type of starter . . . .

Amount of starter . . . .

Time at adding starter . . . .

Milk temperature at time of adding starter . . . .

2. RENNETING

Type of rennet . . . .

Amount of rennet . . . .

Time at adding rennet . . . .

Milk temperature at time of adding rennet . . . .

3. CUTTING THE CURD

Size of curds . . . .

Time at cutting curds . . . .

4. COOKING THE CURD

Time of cooking curd . . . .

Temperature at start of cooking . . . .

Temperature at finish of cooking . . . .

5. DRAINING THE CURD

Time of draining . . . .

6. MILLING THE CURD

Time of milling . . . .

7. SALTING THE CURD

Amount of salt added . . . .

Type of herbs added . . . .

Amount of herbs added . . . .

8. PRESSING THE CURD

Time at start of pressing . . . .

Amount of pressure at start . . . .

Amount of pressure at end . . . .

Date at end of pressing . . . .

9. AIR-DRYING

Date started . . . .

Date finished . . . .

10. WAXING

Date waxed . . . .

11. AGING

Temperature during aging . . . .

12. EATING

Date of first bite . . . .

COMMENTS AND OBSERVATIONS

. . . .

. . . .

. . . .

RESOURCES

Resources for Cheese-Making Supplies

The following companies carry a complete line of cheese-making supplies, including rennet and cheese-making kits.

Beer & Wine Hobby
800-523-5423
www.beer-wine.com

Caprine Supply
800-646-7736
www.caprinesupply.com

Glengarry Cheesemaking & Dairy Supply
888-816-0903
www.glengarrycheesemaking.on.ca

Hoegger Supply Company
800-221-4628
www.hoeggerfarmyard.com

The Home Wine, Beer, and Cheesemaking Shop
818-884-8586
www.homebeerwinecheese.com

Lehman's
888-438-5346
www.lehmans.com

Moorlands Cheesemakers
+44-0196-335-0634
www.cheesemaking.co.uk

New England Cheesemaking Supply Company
413-397-2012
https://cheesemaking.com

COMMERCIAL SUPPLIES AND EQUIPMENT

Alliance Pastorale
+33-05-49-83-30-40
www.alliance-elevage-export.com/en

Cheese-Making Media

Cheesemaking 101 with Ricki Carroll
New England Cheesemaking Supply Company
413-628-3808
www.cheesemaking.com/
    CheeseMaking101DVD.html
Farmhouse Cheddar, Fromage, Blanc, Crème Fraîche, Queso Blanco, Ricotta, Mozzarella, Mascarpone

Groups of Interest

Slow Food
718-260-8000
www.slowfood.com

The American Cheese Society
720-328-2788
www.cheesesociety.org

The Dairy Practices Council
607-347-4276
www.dairypc.org

State Regulatory Agencies

To find information on state agencies that regulate dairy farms and cheese plants and the requirements for a farmstead cheese dairy, go to www.fda.gov/Food/Guidance Regulation/FederalStateFood Programs/ucm114736.

You can ask for information by state.

Websites with Marketing Information

Agricultural Marketing Service
United States Department of Agriculture
www.ams.usda.gov

Growing for Market
www.growingformarket.com

ALPHABETICAL LIST OF CHEESE RECIPES

*These advanced recipes were developed by Jim Wallace, "The Tek Guy" at New England Cheesemaking Supply Company, and have been adapted for this book. You will find stories, in-depth details, and step-by-step photos for all these recipes at cheesemaking.com. Since joining our cheese-making family, Jim has traveled extensively, visiting cheese makers, teaching workshops, and creating recipes for home cheese makers, taking many to the next level.

INDEX

Page numbers in *italic* indicate photos or illustrations; numbers in **bold** indicate charts.

red bacteria (*Brevibacterium linens*), 22, 117
red bacteria ripened cheeses, 126–135, *126–127*
    Brick Cheese, 128–130
    Limburger, *126*, 131–132
    Muenster, *126–127*, 134–135
Red Barn Family Farms, 219
Red-Pepper Spread, Greek, 309
rennet/renneting, 1, 12–14, 39–40
    adding rennet, 40
    animal rennet, 12
    coagulation and, 39
    diluting, 38, *38*, 39
    Junket, 14
    in the old days, 14
    selecting, 12
    thistle rennet, 13
    top-stirring, 40, *40*
    vegetable rennet, 12–13
Ribollita with Parmesan Broth, 319
ricotta, *61*
    Beet Tartlets with Ricotta and Blue Cheese, 312–313
    Galette, Plum and Ricotta, 339–340
    Gnudi with Brown Butter and Sage, 328–329, *330*
    Hotcakes with Lemon Curd, Ricotta, *298*, 299
    Ice Cream, Ricotta Rum-Raisin, 346
    Ricotta, Pure Whey, 255
    Ricotta, Whole-Milk, 82, *83*
    Ricotta with Whey and Milk, *256*, 257

Ricotta Salata, *169*, 179
    Acharuli Khachapuri, 303–304
rind development, 22
ripened cheeses. *See* soft and semi-soft ripened cheeses
ripening cheese, 36–37
    direct heating, 36, *36*
    indirect heating, *36*, 37
Roasted Agrodolce Cauliflower Salad with Goat Cheese, 322
Robiola, 106–107
Romano, 182–183
Roquefort cheese, 224
    Roquefort caves and, 118, *118*

S

Saag Paneer (Spinach with Paneer), 331
Sabot, Topher, 270
Sage, Gnudi with Brown Butter and, 328–329, *330*
Sage Cheddar, 146
Sainte-Maure de Touraine, 245–246, *245*, *250*
Salad Dressings, Buttermilk, 324–325
salads
    Brussels Sprouts and Apple Salad with Smoked Gouda, 324
    Farro Salad with Crème-Fraîche Dressing, 322
    Fatoush with Halloumi Croutons, 319–320
    Kale Salad with Butternut Squash and Manchego, 320–322, *321*

Roasted Agrodolce Cauliflower Salad with Goat Cheese, 323
Saleh, Imran, 94, *94*
salt, cheese, 19
salting, 45–47
    brining, 46–47
    direct, 46
Sandesh, *338*, 339
São Jorge, 192–193
Savory Fromage Blanc Dips, 308
Schav, 318
Schiz, 84
Schumann, Maria, 139, *139*
Seasons Dairy, 114
semi-soft cheese. *See* soft and semi-soft ripened cheeses
Sequatchie Cove Creamery, 167
shaving cheeses, 306, *306*
Shrimp Saganaki (Shrimp Baked with Feta and Tomatoes), 336, *337*
Shropshire, *232*, 233–235
Skillet Pizza with Greens and Prosciutto, *326*, 327–328
skimmer/ladle, *26*, 27–28
"slipskin," 22
smoking cheese, 53, *53*
soft and semi-soft ripened cheeses, 115–117
    aging, 51, 52
    cutting, 280–281, *280*, *281*
    draining, 43
    ladling and slicing, 42, *42*
Sorrel Soup (Schav) with Buttermilk and Whey, 318
Soufflé, Foolproof Cheese, 328–329

ACKNOWLEDGMENTS

A special thank-you to Lisa Hiley and all the fabulous staff at Storey
who held my hand all along the "whey." And to my son-in-law Mark Chrabascz for his
patience and knowledge of recipes and editing.

And to Jim Wallace, a.k.a. The Tek Guy, who has helped us grow in many ways.

KEEP YOUR CREATIVITY COOKING
WITH MORE STOREY BOOKS

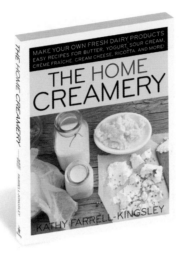

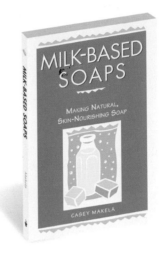

BY RICKI CARROLL & SARAH CARROLL

Step-by-step directions teach how to make 10 cheeses kids love, including mozzarella, cream cheese, and feta. Easy recipes and lively serving suggestions will help you use your handmade creations in dips, spreads, sandwiches, and sweets.

BY KATHY FARRELL-KINGSLEY

Make your own dairy products with these easy methods for butter, yogurt, sour cream, and more. Step-by-step instructions are augmented by 75 delicious recipes that use your freshly made dairy, from Apple Coffee Cake with Caramel Glaze to Zucchini Triangles.

BY CASEY MAKELA

Blend your own moisturizing and nourishing soaps at home using cow's or goat's milk. In addition to 12 specialty bar formulas, you'll find recipes for bath milks and cleansing creams, plus packaging and marketing tips.

JOIN THE CONVERSATION. *Share your experience with this book, learn more about Storey Publishing's authors, and read original essays and book excerpts at storey.com. Look for our books wherever quality books are sold or call 800-441-5700.*